SpringerBriefs in Applied Sciences and Technology

SpringerBriefs present concise summaries of cutting-edge research and practical applications across a wide spectrum of fields. Featuring compact volumes of 50 to 125 pages, the series covers a range of content from professional to academic.

Typical publications can be:

- A timely report of state-of-the art methods
- An introduction to or a manual for the application of mathematical or computer techniques
- A bridge between new research results, as published in journal articles
- A snapshot of a hot or emerging topic
- An in-depth case study
- A presentation of core concepts that students must understand in order to make independent contributions

SpringerBriefs are characterized by fast, global electronic dissemination, standard publishing contracts, standardized manuscript preparation and formatting guidelines, and expedited production schedules.

On the one hand, **SpringerBriefs in Applied Sciences and Technology** are devoted to the publication of fundamentals and applications within the different classical engineering disciplines as well as in interdisciplinary fields that recently emerged between these areas. On the other hand, as the boundary separating fundamental research and applied technology is more and more dissolving, this series is particularly open to trans-disciplinary topics between fundamental science and engineering.

Indexed by EI-Compendex, SCOPUS and Springerlink.

Ahmet Atalay · Biruta Švagždienė

Achieving Environmental Sustainability in the Sports Sector

Nature Means That Future

Springer

Ahmet Atalay
School of Physical Education and Sport
Ardahan University
Ardahan, Türkiye

Biruta Švagždienė
Department of Sport and Tourism Management
Lithuanian Sport University
Kaunas, Lithuania

ISSN 2191-530X ISSN 2191-5318 (electronic)
SpringerBriefs in Applied Sciences and Technology
ISBN 978-3-031-37400-5 ISBN 978-3-031-37401-2 (eBook)
https://doi.org/10.1007/978-3-031-37401-2

This Springer imprint is published by the registered company Springer Nature Switzerland AG
The registered company address is: Gewerbestrasse 11, 6330 Cham, Switzerland

Preface

Climate change and global warming are among the most important problems that humanity has faced recently. The natural environment is exposed to negative effects depending on each human activity. Depending on the production and consumption activities in many sectors such as transportation, tourism and production, greenhouse gas and carbon emissions are increasing day by day, and environmental concerns are gradually increasing. Recently, these environmental concerns have been associated with the sports sector, and the negative environmental effects caused by mass human movements in the sector have been the subject of scientific research.

This study has been prepared with the aim of creating a difference and awareness in terms of revealing the relationship between sports and the environment. In the first part of the book, 17 sustainable development goals put forward by the United Nations are discussed in a comprehensive way in the light of the existing literature. In the second part, the relationship between environmental sustainability and sports has been revealed. The negative environmental impacts originating from the sports sector have been comprehensively conveyed to the reader. Then, examples of measures and practices taken in the sports sector on a global scale for environmental sustainability targets are given.

In the third chapter, the legal measures prepared and implemented for the protection of the natural environment and sustainable environmental goals in Lithuania, the country where the research was carried out, were discussed in detail, and the laws, regulations, plans, policies and strategies that were put into practice more specifically for the sports sector were evaluated.

In the fourth and last part, suggestions have been presented by the researchers in order to reduce the environmental impacts in the sports sector and to reduce the carbon footprint, with the aim of applying it to sports institutions and events.

It is known that more theoretical and practical efforts are required for the bond between sports and environment to be permanent and for sustainable sports and environment. Great responsibilities fall on all sectors and all humanity in order to protect the planet we live on and transfer it to future generations. With the hope that this book, which has been prepared with the aim of making a small contribution to this responsibility, will find its answer in theory and practice.

Ardahan, Turkey Ahmet Atalay
Kaunas, Lithuania Biruta Švagždienė

Acknowledgments This book has been prepared in the light of the data and evaluations obtained in the project titled 'Development of Decision Support System for Carbon Friendly Sports Management' supported by the 'Turkish Scientific and Technological Research Council (TUBITAK) 2219 International Postdoctoral Research Fellowship Program for Turkish Citizens' and carried out by Ahmet Atalay.

Contents

Chapter 1
Sustainable Development

Sustainability has been defined in many different ways in the literature. However, the common point of all definitions is that they have financial, environmental and social dimensions. Sustainability, which basically includes three sub-dimensions, emphasizes the relationship between human life and the living environment. According to the official definition of sustainability made by the United Nations, "sustainability is the protection and transfer of resources necessary for the vital activities of future generations, while meeting the human needs of the present" [85]. In parallel with meeting the needs of the current social life, the planned use of natural resources without interrupting the vital activities of future generations can be considered as basic foresight.

Sustainable development can be expressed as the quantitative and qualitative improvement of all social activities such as education, nutrition and living in better conditions for all living organisms from individuals to society. It is also known that all institutions and organizations have a role in this development and transformation. The concept of sustainability can be defined in many different ways, and definitions can be multiplied in various fields. However, in general, the concept of sustainability, as defined in the Bruntland Report prepared by "the United Nations Commission on Environment and Development" in 1987, is perceived as the ability of ecology and ecological systems to continue their functions and processes [13].

Especially, the developing technology and the increasing use of natural resources accordingly have brought the emergence of environmental problems and the depletion of natural resources to the agenda. The resource scarcity resulting from the unlimited use of natural resources and the negative externality caused by production processes have paved the way for the concept of sustainability. From this point of view, sustainability can only be realized in a way that depends on the use of the resources existing in nature by giving them the opportunity to renew themselves [91].

The environmental problems that emerged nationally and locally during and after the Second World War were ignored in the name of development and enrichment and were considered as a consequence of development that had to be endured. The

A. Atalay and B. Švagždienė, *Achieving Environmental Sustainability in the Sports Sector*, SpringerBriefs in Applied Sciences and Technology,
https://doi.org/10.1007/978-3-031-37401-2_1

fact that development-based activities caused environmental problems did not find an agenda as an issue that needed to be questioned for that period. Although the environmental problems that emerged after the development was achieved were intervened with therapeutic methods and tried to be improved, the interventions made after this situation, not before, did not work and played a role as a factor that dragged the situation to more dire points. The use of natural resources, which has turned into an ever-growing avalanche, and the capitalist economic order that does not allow time to renew itself have caused environmental problems to grow continuously and to be addressed on the international platform [35].

In 2015, the leaders of countries from all over the world came together at the United Nations Headquarters and announced to the world public opinion the 17-article Sustainable Development Goals, which can ensure the common welfare of humanity to eliminate poverty in all its dimensions by 2030. In these articles, which draw a new global development framework, a development agenda has been established, including sustainable cities, climate change, environmental issues such as combating drought and protecting biodiversity and many other issues [59].

"The Sustainable Development Goals" (SDGs) defined by "the United Nations" can be considered as a consensus that emphasizes the theme of a "common future" put forward by the United Nations, aiming to make the present more liveable and at the same time to protect the future. Meadows et al. [47] state that the United Nations primarily addressed sustainable development goals under two headings: "permanent economic growth" and "protection of the environment and natural resources". However, with the decision taken at the World Summit on Social Development in 1995, a third title was added to the sustainable development process as "securing social development" [87]. With the decision taken in 1995, sustainable development was addressed under three main headings. These are economic, social and environmental sustainable development goals. Today, these three topics are handled sensitively for sustainable societies, and societies and countries aim to secure their future while trying to implement their current development goals.

1.1 Sustainable Development Goals[1]

The Sustainable Development Goals (SDGs) are a set of 17 global goals adopted by the United Nations in 2015 as part of the 2030 Agenda for Sustainable Development. The SDGs are designed to be a blueprint for ending poverty (Economic Sustainability), protecting the planet (Environmental Sustainability) and ensuring that all people live in peace and prosperity (Social Sustainability).

[1] 17 targets set by the United Nations are presented to the reader in this section. Direct quoting has been avoided in the transfer of these goals and the goals have been tried to be presented together with the author's comments. Sustainable Development Goals are quoted from the official website of the Department of Economic and Social Affairs. In order not to tire the reader, references were not made separately for each target.

Each goal has specific targets and indicators to measure progress. The SDGs are intended to be integrated and indivisible, meaning that progress on one goal depends on progress on others. For example, achieving gender equality and empowering all women and girls is essential to ending poverty and hunger and ensuring universal access to education and health care.

The SDGs are the responsibility not only of governments but of all actors in society, including the private sector, civil society and individuals. The SDGs provide a framework for all actors to contribute to sustainable development and hold each other accountable for progress. The SDGs are a call to action for all countries and all stakeholders to work together to build a better future for all. They represent an opportunity to end poverty, protect the planet and ensure that all people live in peace and prosperity. The Sustainable Development Goals provide the roadmap needed to create a sustainable world for all.

1.1.1 No Poverty (End Poverty in All Its Forms Everywhere)

Poverty is not new, but it arouses a lot of scepticism every time it is mentioned. This is because it has a very destructive effect on poor people. Poverty reduces people's social and psychological prestige. Poverty is the state of being poor. This can be evidenced even in abundance, because there can be malodorous poverty as a result of a lack of knowledge to translate potential into practical creative benefits to mobilize prosperity. So if there is poverty in something, then there is a lack of it or its quality is extremely low [3]. Therefore, poverty has been positioned as a major target in the last two global development agendas [57].

Eradicating poverty is one of the most pressing issues facing the world today. According to the World Bank, in 2019 more than 736 million people lived in extreme poverty (living on less than $1.90 a day). The first of the Sustainable Development Goals, No Poverty (SDG 1), means ending all forms of poverty everywhere. This goal has two sub-goals: To reduce the proportion of people living in poverty to less than 3% by 2030 and to ensure that all people have the economic, social and political means to end poverty. Ending poverty requires a comprehensive approach that addresses multiple and interrelated causes of poverty, such as *lack of access to education and health services, discrimination and lack of economic opportunities* [86].

According to the United Nations Sustainable Development Goals 2022 Report, the concern about poverty on a global scale is increasing. Especially, COVID-19, which has turned into a global health crisis, and the Russia-Ukraine war are among the determinants of this situation, because according to the report, for the first time in the last 20 years, the working poverty rate increased in 2022. Approximately 8 million people are struggling with poverty and have difficulty in maintaining their economically vital activities despite having a job [86]. Between 2015 and 2018, the global poverty rate continued its historic decline, falling from 10.1% in 2015 to 8.6% in 2018. Today, estimates show that the global poverty rate has risen sharply from 8.3% due to the COVID-19 pandemic, 9.2% in 2019 and 9.2% in 2020, representing the first increase in extreme poverty since 1998 and the largest increase since 1990, delaying poverty reduction by nearly three years [84]. However, another dimension of poverty is people who are not included in the formal financial system. That is, despite progress over the past three decades, three billion poor and more than two billion financially excluded people worldwide live in developing countries [31]. In Africa, for example, only about 4% of the population has a bank account [76], while in Latin America 65% of the population is unbanked [90]. Although poverty is a general problem in societies, the most striking part is that children and individuals under the age of 18 are mostly affected by this poverty. As of 2013, it is quite remarkable that half (385) million of the 767 million extremely poor people are under the age of 18 [53].

It is possible to say that people whose economic assets are under threat despite working and having a job accelerate the global poverty index. In order to combat poverty on a global scale, the United Nations has summarized the policies to be followed under the title of zero poverty in 6 steps:

- By 2030, to at least halve all groups regardless of men, women and children identified as poor through a comprehensive study.
- Carry out nationally suitable social conservation systems and measures for all, embracing all social strata, and meet the defence needs of those in need and poor by 2030.
- By 2030, to provide that all group in society, especially the defenceless and poor, have reaching to base services, land and other shapes of ownership, heritage, "natural resources", "new technology" and "economic services".
- By 2030, increase the resilience of the defenceless and poor and mitigate their exposure and sensibility to Processes related to the climate problem and environmental, social and economic problems.
- To prepare modern and up-to-date development actions and to put these actions into practice with the aim of carrying out plans, policies and actions in a healthy way within the scope of fighting against poverty for developing countries in the world.
- Developing global-scale strategies based on gender-based development plans that include positive discrimination to support investments to end poverty and poverty.

The fight against poverty is not only addressed through economic activities and measures. The goal of poverty alleviation, which encompasses a wide range of issues, including gender equality, access to education, catastrophic climate and natural events and social protection programs, emphasizes that all societies should have the opportunity to live in humane conditions. As societies, countries and governments strive to develop social and economic partnerships so that every individual can live in equal economic conditions, it can also be said that investments aim to achieve economic balance.

Creating humane living conditions and increasing the level of individual and social welfare is one of the main objectives of the fight against poverty. Raising the minimum living conditions, enabling people to benefit from services equally and preventing the formation of privileged classes within the society are essential. The fact that each citizen can live in equal conditions and benefit equally from the services provided by governments is one of the indicators of the elimination of poverty. Especially, the protection of children and young people from the negative effects of poverty makes it inevitable to take concrete and solution-oriented steps in the fight against poverty in order to build a healthy and secure future. In this way, it seems possible to build strong societies through strong individuals.

1.1.2 Zero Hunger (End Hunger, Achieve Food Security and Improved Nutrition and Promote Sustainable Agriculture)

"The United Nations Sustainable Development Goals" aim to reduce hunger to zero in all societies by 2030 and to find a solution to the problem of poverty in all its dimensions. Because the most fundamental problem facing today's societies is hunger and poverty [20]. According to United Nations data, 1 out of every 10 people suffer from hunger, while 1 out of every 3 people cannot meet their nutritional needs regularly. It is stated that COVID-19, conflicts and the Russia-Ukraine war are the determinants of this situation. The UN, which reveals that especially the war conditions affect access to food, reveals that a global hunger crisis has emerged as the war disrupted grain shipments [85].

While hunger and access to regular food, which are among the main problems of humanity on a global scale, are based on sustainable agricultural policies, safe and continuous food access policies are considered as very important topics. In this context, in order to end hunger, ensure access to clean and regular food and eliminate hunger, the UN has summarized the policies to be followed under the title of zero hunger in 8 steps:

- By 2030, to solve the problem of hunger and to ensure that all individuals, especially babies, have access to food and nutrition resources that will help them survive.
- By 2030, to address all nutritional deficiencies, including achieving global goals to combat child ailments such as underdevelopment and underweight, and addressing and solving nutritional and food problems, particularly girls entering puberty, pregnant women and women with children and the elderly.
- By 2030, to double the agrarian productivity and inflows of small-scale food directors, primarily women, residents, families, shepherds and fishermen, through secure and indifferent access to land, other productive coffers and inputs, information and fiscal services, and to produce requests and openings to generate value addition and non-agricultural employment.

- By 2030, to produce and implement plans and policies that create sustainable understanding and programs in food production and distribution, develop environmentally friendly systems, combat climate change and extreme weather conditions, increase the fight against natural disasters and increase the productivity of land and soil.
- By 2020, to protect and maintain the genetic diversity of animals (domestic and wild), seeds, wild species and all plants, including different plant and seed warehouses, managed according to a comprehensive and quality management approach on a global scale, and ensure and encourage balanced, fair and equitable access to all these.
- To create global partnerships and collaborations with the aim of increasing production capacities in agricultural products, increase the agricultural potential of developing countries, contribute to the development of plant and animal husbandry-related sectors and make necessary investments in animal husbandry and gene studies.
- Depending on the Doha Development Round practices, it is aimed to eliminate the confusion and distortions related to the agriculture sector and trade and to take the necessary steps to correct this process.
- To take measures to ensure that all processes in the food sector can function properly, ensure that all information about the sector is accessed accurately and in a timely manner and help combat price imbalance in the industry.

The UN, which has prepared a production-based roadmap especially in the agricultural sector, aims to build the self-sufficient societies through agriculture. In particular, the continuity of access to food is accepted as equivalent to the continuity of production. While it is emphasized that productivity in agriculture will increase, thanks to the protection of ecosystems, especially children's access to food for clean and adequate nutrition is also handled sensitively. While the efforts of social structures and governments towards people's right to adequate nutrition and clean food are concretized with these steps, it can also be stated that individual and social development is tried to pave the way. Regarding this process, the United Nations Information Platform sets out important targets for sustainability in agriculture and food production. According to this, "*as the human population on a global scale increases at this rate and continuously, much more effort and innovation will be urgently needed to sustainably increase agricultural production, improve the global supply chain, mitigate food losses and waste, and ensure access to food for all those suffering from hunger and malnutrition. Many in the international community believe that it is possible to eliminate hunger for future generations and continue their efforts for global co-operation to achieve this goal*" has been strongly emphasized in recent years (United Nations Knowledge Platform).

1.1.3 Good Health and Well-Being (Ensure Healthy Lives and Promote Well-Being for All at All Ages)

The SDG on good health and well-being is a particular challenge at the moment, given the global health threats posed by serious pandemics, including the recent COVID-19 pandemic, which could disrupt national and international progress towards achieving the 2030 goals [36, 42]. Emerging pandemics, conflicts, climate change and global health challenges imposed by weak and inadequate economic growth pose threats to the functioning of the third goal, which focuses specifically on good health and well-being [70]. The UN emphasizes the importance of improving individual and collective health on a global scale. Indeed, the economic consequences of health and well-being can also increase the budget burden of governments. For this reason, the UN has summarized the policies to be followed under the heading of ***health and well-being*** in 13 steps to improve individual and social health and well-being:

- By 2030, to mitigate maternal mortality rates below 70% on a global scale.
- By 2030, to end preventable deaths of new-born babies and children under 5 years of age, aiming to mitigate the new-born mortality rate to at least 12 per 1,000 live births and the under-5 mortality rate to at least 25 per 1,000 live births in all countries.
- By 2030, the Elimination of epidemic diseases that threaten societies, especially AIDS, and the strict fight against these diseases.
- By 2030, to reduce diseases (non-communicable) that cause premature deaths and improve individual and community health conditions.
- Strengthening the prevention and treatment of substance abuse, including the use of narcotic drugs and the harmful use of alcohol.
- By 2020, to mitigate the number of dead and injured due to traffic accidents worldwide as much as possible.
- By 2030, to provide that all segments of society have access to family planning, education and awareness studies on sexual health and reproductive problems and to provide that these practices are included in plans and policies on a country basis.
- Widespread access to quality health services with reduced economic and financial risks; however, to build comprehensive health systems such as access to health services, vaccines, medical supplies and medicines that balance price and quality.
- By 2030, the number of patients and deaths due to pollution by harmful chemicals (soil, air and water) will be significantly decreased.
- Performing global-scale inspections and controls on tobacco and tobacco products according to the relevant criteria of the World Health Organization.
- To endorse the production, development and use of medical materials, drugs and vaccines for all diseases all over the world, especially in developing countries.
- Increasing employment in the health sector on a global scale, especially in developing countries, and expanding and increasing activities for the training and development of existing personnel.

- It is aimed to strengthen the health systems in order to minimize the local and global risks in the health system, manage these risks and reduce their effects all over the world, especially in developing countries.

The key to individual and social health and well-being is the construction of strong and adequate health systems. Countries and governments should focus on the necessary infrastructure and superstructure works for citizens to benefit from health services at an adequate and equal level. In particular, access to services and medicines should be removed from political and economic concerns and should serve health and well-being. Health, which is one of the most fundamental problems of underdeveloped or developing countries, should be one of the priority areas of government policies. It can be said that the United Nations has a proactive understanding of eliminating health problems and improving well-being and expects countries to act with the same understanding.

1.1.4 Quality Education (Ensure Inclusive and Equitable Quality Education and Promote Lifelong Learning Opportunities for All)

The UN goal of quality education can be regarded as a regulation on the right of individuals to equal and quality education and represents the multidimensionality of social development, because it is known that receiving quality education leads to the emergence of people with high skills, which in turn enables them to work for good wages. Therefore, it is shown that individuals whose economic welfare level increases can be healthier and take a more active part in social life [46, 63]. In addition, it can be said that there is a strong relationship between high rates of participation in education in societies and the development of industrial development and innovative studies [9]. Improving countries' understanding of education and increasing its quality has a multifaceted function. It can be said that quality education, which supports other development goals, is one of the determinants of individual and social development. Therefore, the UN has summarized the policies to be followed under the title of quality in education in 10 steps in order to create and develop a quality education understanding in countries:

- By 2030, to provide that all children, regardless of boy or girl, complete their education processes in an effective, qualified, equal, fair and most importantly free way.
- By 2030, to provide that all children, regardless of boy or girl, have easy and convenient access to early childhood development processes and pre-school education.
- By 2030, to assure that all individuals receive quality education, regardless of gender, up to the university, which covers the entire education process.

- By 2030, to increase the digital assets of all individuals, regardless of young or old, who have professional skills and qualifications in employment processes.
- By 2030, to provide that gender-based unequal and disadvantaged groups reach all educational processes in a balanced and equal manner at all levels.
- By 2030, to ensure that all young people and a significant proportion of adults, both male and female, acquire literacy and numeracy skills.
- By 2030, to support the understanding of sustainable societies that include concepts such as human rights, gender equality, war, violence and peace and to support the dissemination of global citizenship understanding.
- To support the construction of safe, effective and inclusive education and training environments for all disadvantaged groups, especially children and people with disabilities.
- By 2030, starting from developing countries, it should be aimed to increase scholarship opportunities and opportunities for entry to university education. In this way, support should be provided to countries where educational activities are disrupted, especially African countries. It should be aimed to increase the quality of education in fields such as computer and engineering, especially in vocational education.
- By 2030, teacher training programs should be supported to increase the quality of education. Global partnerships and collaborations should be established for this purpose.

Education is of key importance for all social developments. The provision of qualified and high-quality educational environments and services is inevitable for progress. It is necessary to develop and extend educational opportunities without discriminating between children, young, old, men and women. Especially in underdeveloped and developing countries, improving the infrastructure and superstructure conditions related to education, and improving the quantity and qualifications of teachers lead the way for quality education. At the same time, the accessibility of education services and the balance between public and private education institutions prevent the formation of social differences. Providing equal opportunities to all social segments within the education system and accepting social status qualities such as gender and income distribution as equivalent to education are among the main determinants of social development.

1.1.5 Gender Equality (Achieve Gender Equality and Empower All Women and Girls)

Gender equality is still a debated issue in modern societies. In all forms of social life, male–female dominance is frequently addressed both in theory and in practice as a complexity. It is essential to eliminate all gender-based uncertainties until women can take part in education, business and social life; girls can benefit from equal opportunities in education, and women are employed at the same level and with the same wage policies as men in business life. In the process that continues from the individual to the society, the evaluation of all actions independently of gender has been on the agenda frequently recently. "Gender equality, which is defined as women and men having equal rights and opportunities in all areas of life (economic, social, political, cultural), is one of the most fundamental rights of human beings. If those who are discriminated against due to gender differences cannot benefit from these opportunities, gender inequality comes to light and becomes one of the most discussed issues" [23]. While emphasizing that strong societies are built through men and women, the United Nations has gathered the policies to be followed under the title of gender equality in 9 steps:

- Stop and reduce discrimination practices, especially against girls and women, in all social settings.
- It should be aimed to prevent abusive and criminal behaviours, especially among women and girls. In addition, plans and policies should be prepared to prevent human rights violations such as human trafficking, sexual assault and abuse.
- It should be aimed to prevent practices that threaten physical and mental health, such as marriage at very early ages and circumcision of women, regardless of male or female. In this way, it can be ensured that children can take more part in social life and that their education processes continue without interruption. Children and youth should be protected from all these inhumane practices.
- It should be aimed to increase and strengthen the infrastructure services provided by state institutions and protection policies. However, responsibilities in family home life should be encouraged. In addition, it is inevitable to define housework clearly and to give the necessary value to these works. Developing and sharing a sense of responsibility is a very important issue as it will contribute to the development of social consciousness.
- It should be ensured that women are included in all processes of social life. The presence of women in education, policy and financial processes should be increased. In this process, it is very important to ensure equality of opportunity for women. In this way, important steps can be taken in the fight against gender inequality.
- Education and awareness of reproductive and sexual health issues should be increased. In particular, the applicability of international agreements should be increased (International Conference on Population and Development Action Program and Beijing Platform for Action).
- It should be ensured that women have property rights like men and have equal rights with men in this regard. In particular, access to financial instruments and resources, use and possession of all property rights will have a positive effect on women's presence in the economic system. Women's equal access to financial life and property rights with men will strengthen women's voices.
- Information and communication technologies are becoming more and more widespread. It should be aimed to use incentives for women's access to and use of these tools. As a matter of fact, information technologies are becoming more

and more widespread in today's modern societies. Through educational activities, women should be provided with the opportunity to use these tools frequently and intensively, and opportunities should be offered.

- It should be aimed to ensure gender equality in societies. Women's presence should be increased at every point of education, trade, politics and social life, and plans, policies and strategies should be developed and implemented for this.

Even today, women are subjected to discrimination, violence and abuse in many societies simply because of their gender. They experience the pains of existence not only as a problem of existence in work, education or social life but also as an individual. Sexist discrimination and crimes, which are sometimes ignored on a global scale, can occur even in many developed countries and societies. In developing countries, while the early or forced marriage of girls is recognized as an important social problem, the processes related to their education are ignored.

1.1.6 Clean Water and Sanitation (Ensure Availability and Sustainable Management of Water and Sanitation for All)

The sixth of the sustainable development goals is related to clean water resources, use and sanitation. It focuses on access transportation processes for the availability and sanitation of clean water resources for all communities. As with all other goals, it focuses on improving the vital processes of all people living on the planet [75]. The sustainability of human life depends on clean water resources. The sixth goal, which is based on the fact that there is no life without water, aims to easy access of people to clean water resources and ensures the sustainability of this situation while at the same time focusing on the protection of clean water resources. All literature reviews point to the year 1990 and after. It is stated that since this date, nearly 3 billion people have not had any difficulties in accessing clean water resources. However, it is stated that close to 700 million people have difficulties in accessing clean water. In addition, it is emphasized that nearly 2.5 billion people have significant problems in accessing the most basic services such as toilets, and this need is met in open spaces [77]. Another risk posed by sanitation is the contamination of clean water sources. Inadequate wastewater systems and often poorly designed and poorly maintained septic tanks cause serious health impacts and contamination of surface and groundwater [89].

Access to clean water and protection of water resources is one of the most important standards of human life of the United Nations. For this reason, the United Nations has categorized the steps to be taken regarding the development goal related to clean water and protection of resources under eight headings: The policies to be followed under the heading of Clean Water and Sanitation are summarized in 8 steps:

- Clean water is the most basic type of life and natural resource. Access to clean drinking water should be provided to all segments of the society and correct price policies should be applied in this process. In this way, inequalities in natural resource access can be eliminated, because the most basic human right is to live.

- By 2030, the cleaning and hygiene needs of all social groups and individuals should be met and equal and fair access to sanitation should be supported. However, it is necessary to prevent other people, especially women, disabled groups and girls, from defecating in open spaces. The most important condition of being healthy is cleanliness and hygiene. In this way, human and public health can be protected and improved.
- It should be aimed to reduce or even eliminate problems such as pollution caused by human mobility and the long-term storage of wastes until 2023. In addition, it is necessary to reduce the chemicals that occur due to production activities and to solve the wastewater problem. In this way, it can be ensured that the clean water necessary for human life is protected and its quality is increased. Therefore, human health can be protected by protecting natural resources and keeping waste away from these sources.
- The use of clean water resources is increasing. Therefore, by 2030, it should be aimed to tackle key fundamental problems such as water scarcity and to affect fewer people from this scarcity. For this, measures should be taken regarding the use of water in all areas and sectors, and the planned and programmed use of clean water resources should be encouraged. It should be explained that natural resources are not unlimited, and a balanced use is required, because human needs are unlimited, but natural resources are not unlimited. Usage and consumption should be controlled.
- By 2030, it should be aimed to develop and implement a management approach related to the use of clean water resources at all levels by establishing global cooperation and partnerships. In this way, limited resources can be used in a balanced and controlled manner and their sustainability can be ensured.
- Water resources are the source of the entire system. By 2030, it should be aimed to protect and ensure the sustainability of forests, vegetation and mountainous areas, especially lakes and rivers, from small to large.
- As stated above, the most important natural resource is water. By 2030, it should be aimed to use water efficiently, to harvest water correctly and to expand wastewater systems. In addition, it should be aimed to establish global cooperation and partnerships and to operate these partnerships effectively for the use and recycling of salt water and the use and dissemination of technologies in this field. In this way, resources can be protected and transferred to future generations.
- In addition to global cooperation, small and local communities at the national level should be encouraged and supported to take responsibility for the management, protection and improvement of water resources.

Not only the existence of clean water resources but also the quality and quality of these resources is extremely important in human life and development processes. Intense population growth and consequent mass human mobility accelerated and increased the use of these resources. Therefore, a result such as the reduction and pollution of resources has emerged. In this context, especially the locking of water is accepted as an important risk in terms of human and public health. Therefore, the protection of these resources should be accepted both as a support for the development process and as a precaution for human health [74]. Especially in developing and underdeveloped countries, it is seen that this pollution is increasing rapidly, and water-cleaning processes are disrupted [83]. Today, it is extremely important to protect and purify clean water resources for every country and society. The protection and improvement of human health should be seen as the primary development goal, and plans and policies should be developed in this direction [74].

In developed countries, people's access to safe and clean water sources, sanitation and wastewater services has improved. However, these services are still a major challenge for less developed societies, often showing inequalities among the most vulnerable communities and regions [27]. In fact, contaminants have proliferated and diversified in recent years, which has become a significant problem for both drinking water and wastewater [16]. Water and sanitation, which is considered as an important problem especially for underdeveloped countries and societies, awaits a solution for the protection and improvement of public health. While the United Nations is trying to develop the necessary steps and collaborations for access to clean water on a global scale, the protection of clean water resources is the most important part of this goal.

1.1.7 Affordable and Clean Energy (Ensure Access to Affordable, Reliable, Sustainable and Modern Energy for All)

Energy and energy policies are at the centre of Sustainable Development processes and goals, because all processes from energy production to distribution and use have significant effects on humans and the natural environment. Human health and sustainable environment and energy use represent an important part of development actions [29]. The need for energy is increasing on a global scale, especially with increasing innovative approaches. The increase in urban life, the growth of industrial organizations and the spread of global perception in production and consumption increase energy consumption. Not only access to energy but also access to cheap and clean energy plays an essential role in environmental and economic sustainability. In particular, the fact that energy needs are met from fossil fuel sources threatens environmental sustainability, while the unbalanced pricing policies of energy markets put the financial sustainability of countries and organizations at risk. The policies to be followed under the United Nations for accessible, clean and safe energy in energy supply and demand, which is one of the biggest threats to economic and environmental sustainability, are summarized in 5 steps:

- It should be aimed to make energy services reliable, financially accessible and up to date by 2030 in all energy groups. The quality of energy services is extremely important, especially since the availability of energy directly affects human life.

- Today, energy needs are met from fossil sources, which poses a threat to a clean and sustainable environment. Therefore, by 2030, it should be aimed to move away from fossil sources and to increase and expand the use of clean and renewable energy.
- As in water resources, effective and efficient use of energy should be expanded and the principle of efficiency in energy should be developed until 2030.
- It should be aimed to take the necessary steps by 2030 for the production, supply and use of clean and safe energy. For this, plans and policies can be created to move away from fossil sources, to provide incentives for renewable energy and to use efficient energy. In addition, it is necessary to plan resources and investments for clean energy research and to support research processes with current technology.
- With the plans and policies prepared on a global scale, it should be aimed to strengthen the energy infrastructure in all societies, especially in developing countries, to develop technology in this field and to improve clean energy policies until 2030.

In 2017, the United Nations published a report stating that SDG7, which calls for affordable and clean energy, has been severely hampered and remains below the desired level. The same report emphasizes the importance of providing energy access for all countries and societies and achieving renewable energy and energy efficiency targets [29]. Accessible and clean energy, which is the main source of sustainable societies, is vital for all sectors from production to food, transport to education. Therefore, affordable and accessible energy, which is expressed as the most important part of sustainable development, has a key role in sustainable development.

1.1.8 Good Job and Economic Growth (Promote Sustained, Inclusive and Sustainable Economic Growth, Full and Productive Employment and Decent Work for All)

The most important factor that determines the quality of human life depends on employment opportunities and the right, improved working conditions. The eighth of the development goals sets out the goals and objectives related to employment and working life. In this context, the scope of the eighth development goal has been determined as follows: Increasing the Gross National Product and its balanced distribution, improving working conditions while increasing employment opportunities, employment policies focusing especially on the young population, effective and efficient use of resources in employment, child labour and informal employment and forced labour. It is aimed to focus on sustainable policies, especially in the tourism sector, and finally to reduce negative effects such as supporting financial resources in employment policies [78]. In building sustainable economic structures, it is very important to establish healthy working environments and to prepare labour rights in accordance with human rights. In this way, the development that will occur in the level of welfare is an indicator that can be felt not only by capital owners but also by employees. Decent work, development of employment, combating the problem of child labour and ensuring occupational safety can mediate the improvement of the welfare level of employees. The United Nations has summarized the policies to be followed under the title of decent work and economic development in 12 steps:

- In order for economic and commercial growth to be felt in human and social life, it should be aimed to regulate the financial growth in the countries according to the commercial and economic conditions in the relevant country. In addition, it should be aimed to increase the financial progress and economic growth rates in underdeveloped countries and to achieve an increase of at least 7%. In this way, it may be possible for global trade indicators to affect human life, and economic mobility may affect the level of welfare.
- It should be aimed to strengthen and expand the digital and technological infrastructure, to make improvements in this regard, to provide diversity in production activities and to exhibit innovative approaches within all commercial sectors, especially in sectors with high commercial value and based on manpower. In this way, the labour market can be included in the production processes and draw a more productive profile.
- It is aimed to build environments that are suitable and worthy of human rights and that work in the process of production activities and to support an innovative approach in these environments. Improving working conditions will have a positive effect on productivity. In addition, plans and policies should be produced to support small-, medium- and large-scale enterprises, to grow them and to reach a more productive point.
- In this period of rapid increase in production and consumption, effective and efficient use and consumption of natural resources on a global scale should be supported until 2030, and programs, strategies and policies should be established for sustainable production and consumption processes. Otherwise, unplanned consumption of limited natural resources will result in negative economic, social and environmental effects.
- Until 2030, employment opportunities should be increased in all social groups, including men, women and disabled groups, and equal and fair working processes and wage opportunities should be created in this regard. The only way to build productive and productive societies depends on improving working life and employment opportunities.
- The inclusion of young people in working life and their involvement in production processes should be supported until 2030. For this, it is inevitable to use all educational tools, especially vocational training. Qualified manpower can be created with an educated workforce. This, in turn, can help increase economic well-being.
- It should be aimed to prevent unregistered and forced labour and to improve heavy and bad working conditions. In addition, it should be supported to prevent social problems such as employing children and including children in the military service process. It is necessary to take steps to solve the child labour problem, especially until 2025.

- It should be aimed to protect workers' rights by providing healthy and correct working conditions. Especially in recent years, due to the increase in the number of immigrants with the wars, the problems of unregistered workers and unsecured employment were a significant threat and solutions should be found for this.
- With globalization, the importance of local elements and values is increasing. For this, local production processes and employment opportunities should be supported until 2030, and sustainable practices should be supported, especially within the tourism sector.
- The availability of financial and economic services for all segments of the society should be supported and the business potential of financial institutions should be increased. In this way, economic mobility can spread throughout the society.
- In order for financial mobility on a global scale to spread all over the world, it can be aimed to support financial and commercial formations in all societies, especially in developing countries.
- Considering the Global Jobs Pact, it should be aimed to employ the young population in societies in different sectors by 2020. Especially in the fight against unemployment, employment opportunities for the young population should be increased and supported.

Meeting the increasing and diversifying human needs in the era of development and change has been associated with the quality of life. Therefore, sustainable development processes focus on meeting human needs and increasing the quality of home life [10]. In other words, sustainable development requires that wealth does not comprehensively decline over time [60]. The main problem is not only the decline in wealth but also the imbalance in economic development and welfare. This imbalance is also increasing. In order to ensure that the level of economic prosperity can be felt by all segments and to eliminate inequalities in income distribution, issues such as increasing employment opportunities, improving business environments and training qualified human resources come to the fore.

Within the scope of a study conducted by [41], assessments of SDG 8 of 20 developed countries during 2013–2018 were made. A visual of this assessment is presented (Fig. 1.1).

A closer look at G20 countries' assessments of SDG 8 shows that Japan has achieved the best labour environment and the most significant economic growth in the period 2013–2018.

The United States ranks second and the Republic of Korea ranks third. The last three G20 countries are Argentina, Brazil and South Africa [41]. The fact that countries with a strong social structure are higher in business and economic development rankings can be interpreted as that all processes from production to consumption are carried out in a balanced manner among stakeholders. A good representation of the elements of social balance in business and economic indicators accelerates economic development, and development can be reflected in society in general.

Country	Relative closeness	Ranking
Japan	0.644759	1
United States	0.643069	2
Korea, Rep.	0.613329	3
India	0.612212	4
China	0.609138	5
United Kingdom	0.605965	6
Australia	0.605807	7
Indonesia	0.599021	8
Germany	0.592811	9
France	0.588447	10
Turkey	0.582485	11
European Union	0.580931	12
Canada	0.567808	13
Russian Federation	0.554956	14
Italy	0.544895	15
Mexico	0.524549	16
Saudi Arabia	0.516044	17
Argentina	0.428712	18
Brazil	0.41011	19
South Africa	0.250499	20

Fig. 1.1 Assessments of SDG 8 of 20 developed countries

1.1.9 Industry, Innovation and Infrastructure (Build Resilient Infrastructure, Promote Inclusive and Sustainable Industrialization and Foster Innovation)

The natural resource potential of the planet on which we live is the main guarantor of human life. However, the world population is increasing and it is accepted that more than 8 billion people live. Therefore, careful, planned and controlled use of existing resources is extremely important in terms of sustainability. The United Nations, while presenting its understanding of sustainability, acts with the understanding of meeting the needs of today and transferring them to future generations. Therefore, resource use has an important role in sustainability [19]. With the developing technology, all sectors, especially production and consumption, are experiencing a transformation. Technological infrastructures are being strengthened. Production sector and infrastructure systems increase their equipment in parallel with the developments. Supply-demand balance is tried to be provided on a global scale depending on all progress and developments. Increasing and diversifying human demands and needs are also trying to be met with the global supply chain. In this context, the United Nations has collected the policies to be followed under the title of industry, innovation and infrastructure in 8 steps:

- It should be aimed to support infrastructure works in all vital areas. In particular, an infrastructure culture with fair and equal access and access to all social segments should be built. Starting from the region, these infrastructure elements should be developed continuously for global development moves.
- Industrial development and industrialization should be supported. In this process, sustainability should be put at the centre, and the focus should be on increasing the share of employment in industrialization processes. In this way, economic development can be developed with a focus on both production and employment. Unemployment can be reduced to a certain extent and the welfare level can be increased.
- Access to financial and economic services should be facilitated for small businesses as well as large enterprises to support employment and production. In addition, the commercial presence of small businesses in developing countries should be supported and their participation in international markets should be encouraged. In this way, the production chain is prevented from entering a certain monopoly. The widespread financial structure can be strengthened on a global scale.
- While supporting the effective and efficient use of resources in production and consumption processes with developing societies, it should be aimed to strengthen environmentally friendly technology infrastructure and industrial processes. In this way, a sustainable understanding can be displayed while maintaining the production balance.
- The most important criterion of the modern age is innovation. In this context, it should be aimed to increase research and development activities and to strengthen the technological infrastructure until 2030. Starting from developing countries, scientific research processes and innovative understanding should be supported. Innovative approaches are inevitable to support social productivity. Sustainability and productivity in the modern age are possible with innovative approaches.
- In order to ensure global balance for small and underdeveloped countries, financial technological infrastructure should be strengthened and these countries should be included in sustainable production processes. In order to achieve social equality, it is inevitable to eliminate the imbalance between countries.
- Especially for developing countries, support should be provided for the development of research and development studies and innovative understanding. For this, plans and policies on a global scale should be formed and implemented. The development of underdeveloped countries can be encouraged with realistic and applicable policies.
- Information technologies are developing rapidly. In order to reflect this development to the whole world, support should be provided until 2020 and affordable prices should be made especially for Internet access. The Internet has become the most important tool of the modern age and all societies should be provided with comfortable and easy access.

The ninth sustainable development goal is a crucial goal that recognizes the importance of infrastructure, industry and innovation in promoting sustainable development. Focusing on building resilient infrastructure, promoting inclusive and sustainable industrialization and fostering innovation, this goal aims to boost economic growth, decrease poverty and improve the lives of people around the world. Achieving this goal requires making the necessary investments, creating an enabling environment and adopting a holistic approach to sustainable development.

1.1.10 Reduced Inequalities (Reduce Inequality Within and Among Countries)

The Tenth Sustainable Development Goal aims to decrease inequalities within all societies and between nations. This is because inequalities within societies are considered to be one of the most important obstacles to sustainable development. In order to build a more comprehensive and just world order based on the principle of equality, it is essential to reduce all inequalities and to place social structures on a balanced basis. Inequalities within social structures can be categorized as economic, social and political. Economic inequality refers to the unequal distribution of earnings and income, whereas social inequality is characterized by differences in access to education, health and other basic services. Political inequality refers to unequal representation and participation in the political process.

The most important problem affecting human life is the imbalances and inequalities that occur on a regional or global scale. Even if it is a developed society, inequalities are often striking in this society. Therefore, while there are inequalities between different groups in a single society, there are inequalities between countries, and this situation is growing and continuing [14].

Therefore, the tenth sustainable development goals are based on an equitable and equal use of financial benefits by all segments, because in countries, the economic balance is deteriorating and inequalities become the main problem of all social segments. For this reason, social development, increasing the welfare level, strengthening social security systems and creating funds in this regard will play an important role in the fight against inequalities [30, 55]. In order for all these processes to be carried out in a healthy way, the United Nations has summarized the policies to be followed under the heading of reducing inequalities in line with the sustainable development goals in 10 steps:

- On a global scale, based on the income distribution of countries, an income increase should be achieved for 40% of the total population by 2030. In this conditions, the opportunity to live in better conditions can be created and the quality of life can be improved.

- By 2030, the political, financial and social participation of all segments of society, regardless of women, men, disabled, young, old and different ethnic and religious backgrounds, should be supported and increased. In order to build participatory societies, despite all the differences, the collective consciousness should be supported and increased.
- It should be aimed to take all steps to prevent and eliminate all plans, policies and strategies that have a different and discriminatory content and to support these steps. In this way, social inequality can be eliminated and equality of opportunity can be established.
- Social protection and wage policies, especially financial measures, should be strengthened and steps should be taken to build equal and fair societies.
- It should be aimed to support, strengthen and improve the institutions and organizations that provide financial services globally. Fiscal sustainability can be achieved by creating plans and policies for this.
- It should be aimed to support the administrative structures and decision-making processes of institutions and organizations providing financial services to be transparent and legitimate. In this way, supervision and control mechanisms can be operated more powerfully.
- It should be aimed to carry out migration policies within a certain plan and program and to make legal arrangements for this. In this way, the mobility of individuals can be made safer, and regular migration mobility can be provided.
- Within the scope of WTO practices, special and exceptional practices should be expanded and expanded, especially for developing countries.
- With national plans and policies, it should be aimed to make investments from foreign countries, especially for underdeveloped countries and societies and to apply different financial incentives for this. In this way, a balanced distribution of investments on a global scale can be achieved. In this way, an economic balance can be established.
- It should be aimed to reduce the costs incurred due to money transfers and remittance transactions on a global scale to below 3% by 2030, and to reduce costs above 5%.

The tenth sustainable development goal is a crucial target that recognizes the importance of reducing inequality to create a more equitable and just world. Achieving this goal requires addressing the root causes of inequality, investing in education and health services, and promoting inclusive and sustainable economic growth. By reducing inequality, we can create a world that is more equitable and provides opportunities for all people to realize their full potential. There are unparalleled financial, political, social and environmental structures in some countries and societies around the world and these need to be followed. Despite the social problems that have emerged recently, including COVID-19, it is inevitable to establish more fair and equal social structures across all countries and to develop plans, policies and strategies for this [14].

1.1.11 Sustainable Cities and Communities (Make Cities and Human Settlements Inclusive, Safe, Resilient and Sustainable)

Depending on population growth and technological developments, people have not been satisfied with meeting their basic needs, but have tried to meet their unlimited demands by using limited resources. In this process, the use of natural resources and the increase in the amount of waste have led to the emergence of many environmental problems. Depending on the rapid population growth, the need for food has increased and fertile agricultural areas have been destroyed for the sake of urbanization in order to meet the increasing needs, and the problem of unplanned urbanization in cities with migration to cities has accelerated the deterioration of the natural balance [2].

Depending on industrialization, a rapid urbanization and an intense increase in the urban population are observed especially in Europe. Depending on the research results, 70% of the population of the European continent lives in cities and this situation is accepted as a necessity of the modern age [69]. Rapid urbanization can of course be considered as a positive trend. The concentration of production facilities and industrial investments in central cities is considered as an indicator of this phenomenon. However, this process may also bring along a number of negativities, for example, overuse of resources, pollution, congestion, irrational consumption patterns, excessive waste production, unemployment and migration.

It is extremely important to be able to combat the above-mentioned negativities in terms of the health and safety of people and the planet. In this context, especially the European Union has placed the goal of sustainable urban development at the centre of its policy agenda for urban areas [67]. The basis of these policies is the concept of sustainable and smart cities. In modern societies, the dissemination of the concept of smart cities can increase the level of financial competitiveness and seems to be an effective and appropriate strategy to ensure the protection and sustainability of the natural environment [43, 68].

In order to combat the problems caused by rapid urbanization, especially in the European continent, the United Nations has included the title of sustainable cities and society among the development goals. Within the scope of the 11th Development Goals of the United Nations, the steps to be taken in line with the objectives of sustainable cities and society are grouped under 10 headings:

- Ensure everyone has access to safe, equitable, balanced and affordable accommodation and simple services by 2030 and improve slums.
- By 2030, accessible travel and transport infrastructure and systems must be provided for all in all societies, such as financial and security. In particular, public transport systems should be expanded and encouraged. In addition, it should be aimed to secure the safe travel rights of women and children, especially individuals with special needs.

- Establishing a sustainable participatory understanding in all cities and urban areas on a global scale; creating plans and policies for environmentally friendly residential areas until 2030.
- Supporting plans, policies and efforts to preserve economic, cultural and natural heritage around the world.
- By 2030, it should be aimed to reduce the number of deaths, to protect the poor and people in need of protection and to minimize the resulting losses by protecting the affected people and societies against natural disasters.
- Due to increasing urbanization, important environmental problems arise. One of them is air pollution and waste. In order to prevent these problems and reduce negative environmental impacts by 2030, cooperation with local governments should be developed on issues such as waste management.
- By 2030, it should be aimed to increase accessible green spaces for the health and welfare of the elderly, women and children, especially the disabled, and to implement the principle of social use in public spaces.
- Urbanization has increased with industrialization. However, with the connection of urban areas with rural areas in societies, it should be aimed to show a balanced distribution of financial development throughout the country. It is extremely important to create plans and policies, especially for the connection between urban and rural.
- A strong risk management approach should be established against increasing natural disasters by 2020. In addition, plans and policies should be established for the creation of comprehensive natural disaster-resistant settlements. Especially, densely populated regions pose significant risks. Necessary steps should be taken to mitigate and reduce these risks.
- In underdeveloped societies, financial and technical support should be provided for the use of local materials in production and construction processes, and it should be aimed to build buildings that are resistant to natural disasters in these countries.

Safe and accessible housing is recognized as a basic human need. Countries on a global scale are trying to adopt the understanding of urban sustainability. The main indicator here is that human capital development is linked to a re-appreciation of healthy living standards and environmental concerns [4, 72]. The joint evaluation of the United Nations' sustainable cities and social health understanding has also been decisive in terms of the development of social capital. However, there are points to be noted here. In addition, there are measures to be taken for a city to be sustainable. Thanks to measures to be taken, sustainable practices in education, buildings, facilities and their management strategies; comprehensive and environmentally friendly waste management can be established and the execution of all these processes conducted in harmony with nature. In addition, the conscious and high level of awareness of the residents of the city is also extremely decisive. Otherwise, it can be said that the measures taken alone will not mean anything [65].

The United Nations has implemented and is following the goal of sustainable cities and societies for liveable and accessible urbanization approach all over the world. As mentioned above, urban sustainability is determinative for the continuity of societies and human health, for increasing human welfare and for the development of social capital. In addition, since the main subject of sustainability goals is human health and development, intense energy consumption, high energy use and waste generation associated with them endanger human health. In addition to taking the necessary measures in cities, education and awareness raising of societies on this issue is another factor. The emphasis is always on human health and community integrity. Therefore, it is important for people to act responsibly and take initiative in all processes that concern them, for the success of the determined goals.

1.1.12 Responsible Consumption and Production (Ensure Sustainable Consumption and Production Patterns)

The production and consumption behaviours of human beings have many environmental effects such as carbon dioxide emission, reduction of biological diversity, causing global warming, and consuming natural resources, directly or indirectly. These effects have been discussed intensively in recent years. Many factors, especially population growth and rapid economic development, play a role in starting this discussion. Some of these factors are

- Globalization and the economies of scale it creates.
- Decrease in product price levels.
- Cultural changes (urbanization, women's entry into business life, etc.).
- Increasing the quality of life of individuals.
- The emergence of environmental problems [34].

There has been a rapid increase in production and consumption due to the increasing population and intensifying human activities. However, this process has brought along intensive resource use, energy consumption and waste production and has become a topic that needs to be taken precautions. The United Nations refers to the sustainability of the production and consumption chain for a sustainable world. In this context, the United Nations has included the title of sustainable consumption and production among its development goals. Within the scope of the United Nations 12th Development Goals, the steps to be taken in line with the objectives of sustainable production and consumption have been gathered under 11 headings:

- Within the framework of the understanding of sustainability in production and consumption, plans and policies for developing countries under the leadership of developed countries should be created. These plans and policies should be included in 10-year development programs.
- By 2030, measures should be taken for the effective, efficient and sustainable use of natural resources on the planet.

- There is a high amount of food wastage due to the excessive consumption process. It should be aimed to reduce this waste in half and to prevent food waste and losses in production and supply processes.
- An environmentally friendly approach should be introduced in the collection and disposal of all waste groups and chemical wastes until 2020. In addition, it should be aimed to reduce the negative effects and pollution caused by waste in a meaningful way.
- Waste production has increased considerably today. In order to reduce and control this, it should be aimed to reduce environmental impacts by creating waste management systems for the reduction, collection, storage and recycling of waste until 2030.
- In particular, the widespread use of sustainability-related practices by international businesses and companies and the continuous reporting of these practices should be supported and encouraged.
- To promote sustainable applications of public institutions in procurement processes within the scope of national plan policies and strategies.
- By 2030, it should be aimed to carry out education and awareness studies and to expand these studies in order to build living standards compatible with the development processes of people and the natural environment in societies. In this way, a nature-friendly life cycle can be established in the development process.
- For a sustainable production and consumption cycle, it should be aimed to construct and build technological infrastructures in developing countries and to strengthen this infrastructure.
- In order to create comprehensive employment, a tourism understanding that supports local culture and values should be built. For this, development indicators should be prepared, and support and incentives should be provided for development. Localization should be given importance and local production processes should be harmonized especially with the tourism sector.
- Tax systems and taxation processes should be modernized and brought up to date. In this process, the conditions of the countries should be taken into account. In particular, fluctuations and wasteful attitudes and behaviours in the markets should be controlled. Steps should be taken to gradually remove subsidies. The relationship of all these processes with the environment should be taken into account. Action plans should be created for the breakdown and development of the poor classes in society.

Sustainable consumption is the way of meeting our present needs without overloading natural resources and destroying the environment, taking into account the possibilities of meeting the needs of future generations [52]. It also uniquely challenges each of the economic sectors—private, public and government. At the household level, buying less in the short term means making significant changes in lifestyle eventually [38]. In today's world, where the problems in production and administrative processes have become more evident, the factors that directly affect production are global problems such as climate crisis, increase in temperatures and drought. Particularly in the sectors related to production, there is an intense use of energy,

Table 1.1 Motivations for sustainable production

Why sustainable production approach?
• Environmental concerns • Decreased non-renewable resources • Stricter legislation • Sustainable production incentives, grant supports • High energy costs • Increasing consumer preference for environmentally friendly products (manufacturing)

and accordingly, significant environmental effects arise. In fact, there is a mutual interaction. While energy consumption increases, the production sector is affected, but, on the other hand, this energy use causes significant damage to the environment and nature. In addition, with the increase in demands due to population growth, the intensity of competition in the sectors increases. Sustainable production motivations put forward to combat the expressed negativities and concerns are presented visually in the Table 1.1 [37].

Sustainability of production, minimum environmental impact, planned use and protection of existing natural resources is a concept that expresses an economic and safe management approach for all people. It is accepted as a process that takes care of the balance with the natural environment while performing production supply [37]. The United Nations focuses on protecting the living resources of future generations by taking steps on a global scale to create a sustainable production and consumption chain. Especially today, the fact that the balance of supply and demand is quite high increases the importance and meaning of this target even more.

1.1.13 Climate Action (Take Urgent Action to Combat Climate Change and Its Impacts)

It is stated that consumption, depending on the increasing human population in today's modern society age, accelerates global climate change by negatively affecting the natural environment [71]. Global climate change, especially in industry and transportation, met the required energy from fossil fuels, the destruction of forests and many other human activities as a result of the greenhouse effect caused by the increase in methane and carbon dioxide gases in the atmosphere as well as synthetic chemicals (chlorofluorocarbon-CFC). It can be defined as the increase in the average temperature [6]. Global climate change is one of the most important environmental problems that arises with the influence of human beings, affects the whole world with its consequences and is widely discussed [18]. Especially in recent years, extraordinary weather events in different geographies of the world, and the rapid melting of glaciers, make the effects and extent of global climate change felt. Although states carry out many studies to reduce the emission of greenhouse gases in cooperation, they cannot be said to be successful enough [5].

The fact that we are in a destructive social period triggers global climate problems. Increasing temperatures, melting glaciers and high carbon emissions are putting the future of humans and the planet in danger. The United Nations associates the fight against global climate change with the existence of human beings. In this context, the fight against the global climate crisis is taken as a basis all over the world. In this context, the steps to be taken in line with the objectives of combating the global climate change problem within the scope of the United Nations 13th Development Goals have been gathered under 5 headings:

- Measures should be taken regarding climate change and natural disasters on a global scale, and the ability of societies to act in harmony in the process of climate change should be strengthened.
- The problem of global warming and climate change is felt all over the world and its effects are increasing. Countries related to this problem should develop plans, policies and strategies at the national level and implement them strictly.
- Education and awareness studies on climate change and global warming problems should be carried out. In this way, it may be possible to reduce the effects of climate change. In this regard, a series of measures should be taken, especially early warning systems.
- The Climate Change Environmental Convention has been accepted by many countries. Under this convention, developed countries should strengthen their actions to reduce the effects of climate change. However, financial support should be provided for developing countries, because combating climate change is the responsibility of all societies. Therefore, it is inevitable for developed countries to take responsibility and provide support to other countries.
- In order to combat climate change on a global scale, all societies must take responsibility. In this context, it should be aimed at countries to prepare plans and policies and to manage this process effective. In order to benefit from the influence of social groups, they need to be encouraged and supported.

In the solution of environmental problems such as global climate change, it is very important to create a conscious society that can be possible by raising individuals who are sensitive to the environment, conscious and actively participate in the solution of problems. Educational institutions, non-governmental organizations and media have important duties to increase social awareness. In addition, the visibility of the activities carried out by public institutions and organizations to reduce the effects of climate change and the awareness of individuals about the issue are other factors that are effective in combating climate change [5].

The problem of climate change, which affects and is affected by all vital activity areas, is accepted as the problem of the entire planet. Many different solutions, especially clean and green energy, are recommended and implemented by the United Nations. It is inevitable to take measures in all sectors from production to agriculture, from energy to sports. Social mobility, especially with increasing human activities, continues to be a source of the problem of climate change. At this point, it is of vital importance to take individual and social measures and strictly implement plans and policies.

1.1.14 *Life Below Water (Conserve and Sustainably Use the Oceans, Seas and Marine Resources for Sustainable Development)*

The source of life is "water". It is an indispensable basic source of life for the continuation of humans and all other living things. And more than 70% of the planet earth we live on is covered with water. The human life, and economic, social and environmental impacts of the oceans and seas are unpredictable. According to the report prepared by the United Nations, the economic value of marine and coastal resources and the industries they create is approximately 3 trillion USD per year, in other words, it is accepted that they constitute approximately 5% of the global GDP [83]. Since both saltwater sources and freshwater sources are accepted as the source of life, it is very important to ensure the continuity of these sources. "Source of life". I think it would be wrong to express the importance and value of the sense in another way. Therefore, one of the 17 development goals of the United Nations is to protect and sustain the ocean, sea and water resources. Within the scope of the 14th Development Goals of the United Nations, the steps to be taken regarding the protection and sustainability of ocean, sea and water resources are gathered under 10 headings:

- Although activities carried out in terrestrial areas may seem irrelevant, they can cause marine pollution. Especially the waste and garbage left in the seas can pose a threat to nutritional health. Therefore, it should be aimed to reduce and even prevent marine pollution in all its aspects by 2025.
- In order to preserve the diversity and productivity of the world's oceans, negative actions that will affect the oceans must be avoided until 2020. For this, positive steps must be taken to properly manage ocean life and protect the ecosystem.
- By establishing science-based partnerships and collaborations in the international arena, steps should be taken to protect ocean life and negative effects should be reduced as much as possible. The oceans are the source of many different kinds of life, and protecting this resource is extremely important for sustainability.
- The biggest threat to ocean life is illegal and uncontrolled fishing methods. It should be aimed to control these hunting activities by 2020. In addition, plans and policies should be developed in order to protect fish diversity and sustain natural life.
- By 2020, it should aim to protect both national and international coastlines and coastlines. These conservation efforts should cover at least 10% of the shorelines. It is extremely important to deal with all these processes with their legal dimensions and to use legal instruments.
- Subsidies for illegal and unregistered fishing are among the most important risks. By 2020, all subsidies for illegal or unregistered fishing activities must be eliminated and strictly prohibited. Subsidies should be made to make poaching more difficult and should be supported by laws.

- Until 2030, especially for the development of developing countries and sustainable economic development, commercial activities based on fisheries, especially fisheries, and all steps towards tourism should be supported and expanded.
- It should be aimed at establishing ocean health and sustainability through scientific cooperation and partnerships on ocean and fishery products. In this way, economic development based on fisheries and aquaculture can be used for developing countries. In this process, the Criteria of the Intergovernmental Oceanographic Commission and the Guidelines for the Transfer of Marine Technology should be taken into account.
- To enable small-scale amateur fishermen to access marine resources and markets.
- It should be aimed to protect and sustainably use the ocean and all resources connected to the ocean on a global scale. The United Nations Convention on the Law of the Sea should be taken into account in the execution of this process.

The oceans, in particular, are an important source of trade and livelihood and have the potential to provide a wide variety of services. It has the potential to create economic value in many different fields, especially in transportation, tourism and fishing. However, it is one of the most important renewable energy sources today. The oceans are also home to the largest connected ecosystem that provides services such as climate stability, oxygen production, nutrient cycling and food. It also makes significant contributions to human welfare through the income it provides [32].

Oceans and seas, which are a balance element for the ecosystem, are a natural solution for economic development as well as increasing construction welfare and climate problems. As the guarantor of life, they are the habitats of thousands of species. In this context, the understanding of sustainable ocean, sea and water resources should be expanded. In addition, benefiting from its economic value at the maximum level can be considered as an opportunity for economic development.

1.1.15 Life on Land (Protect, Restore and Promote Sustainable Use of Terrestrial Ecosystems, Sustainably Manage Forests, Combat Desertification, and Halt and Reverse Land Degradation and Halt Biodiversity Loss)

According to United Nations reports, approximately 10 million hectares of forest areas are destroyed every year due to agricultural activities. The protection and sustainable use of forests, which are an important part of the ecosystem, is very important for human life. "Managing forests sustainably, combating desertification, halting and reversing land degradation and halting biodiversity loss" is one of the priority work and target areas of the United Nations [82]. Sustainable Development Goal 15, Life on Land, identifies biodiversity loss as a key sustainable development problem. As with water resources, the protection of forests is an essential issue for the continuation of life. Within the scope of the United Nations' 15th Development Goal, the steps to be taken regarding the Life on land title have been gathered under 12 headings:

- Terrestrial areas make up a significant portion of the earth's surface. For this, it should be aimed to protect all terrestrial areas containing freshwater resources such as forested areas, mountainous areas and wetlands until 2020. However, the sustainable use of these areas should also be encouraged.
- Forest losses are a huge ecosystem threat. Therefore, efforts should be made to reduce forest losses, prevent deforestation and restore damaged areas by 2020, and afforestation should be given priority.
- With the global warming problem, drought and desertification are increasing rapidly. Therefore, fertile land and soil structure is decreasing. In order to combat this problem, it should be aimed to improve and protect the land structure until 2030.
- Ecosystems can be considered as an important development tool as they are associated with development. In order for the development to be sustainable, mountainous lands and the ecosystems within these lands should be protected until 2030.
- Natural habitats for all living things are an important requirement for biodiversity. By 2020, it should be aimed to take steps and take necessary measures for the protection and improvement of natural habitats.
- Natural and genetic resources are used intensively. It should be ensured that this use is balanced, equal and fair. In particular, it should be aimed to ensure that access to these resources is fair and balanced. Equal and fair use of resources is extremely important for sustainability.
- Termination of commercial and economic activities for protected flora areas; in addition, action plans should be prepared to control commercial activities related to wildlife related to these flora.
- With the aim of protecting ecosystems, it should be aimed to combat invasive alien species and prevent them from damaging ecosystems by 2020. In addition, necessary steps should be taken to eradicate these species.
- There are national local development plans. These plans should aim to integrate values related to biodiversity and ecosystems by 2020. Especially in the fight against poverty, these development plans are of great importance.
- In order to protect ecosystems and biodiversity in nature and to ensure a sustainable natural life, countries should increase their financial resources until 2020 and give more space to this process in their budgets. Since each action will have an economic cost, ecosystem and biodiversity topics should be emphasized in budget planning.
- In order to operate the forests owned by the countries in a sustainable way, it should be aimed to increase the economic tools and financial resources and to use them effectively. In addition, incentives should be applied to increase forest areas for developing countries.

- Steps should be taken to increase sustainable sources of income for small and local communities, primarily aimed at supporting and refining poaching and some species conservation initiatives.

The land balance on the planet is extremely important for human life. The destruction of these lands and the deterioration of their natural balance are considered as threats to human life. Therefore, soil and land management processes need to focus on these issues. For this, it is necessary to produce plans and policies on a national and international scale and to implement strategies. Therefore, the main scope of the 15th development goal focuses on the sustainable use of land and lands [24]. It is known that terrestrial destruction is the most important determinant of climate change and climate shocks. Therefore, it can be stated that the most important fight against climate shocks is to make terrestrial life sustainable. It is stated that desertification and drought problems have arisen more than ever, especially due to the deterioration of terrestrial life and biodiversity [57].

1.1.16 Peace Justice and Strong Institutions (Promote Peaceful and Inclusive Societies for Sustainable Development, Provide Access to Justice for all and Build Effective, Accountable and Inclusive Institutions at all Levels)

It is stated that this goal has been determined in order to create environments where everyone can feel safe, to ensure that the peace environment is permanent, to open all public and private institutions to everyone's access and to make them accountable [85].

The fact that the exploitation of children as a labour force and their exposure to traffic risk has deepened, especially with COVID-19, reveals the necessity of peace and safe environments. According to 2018 data, it is stated that one out of every three people who died due to traffic accidents is a child. According to the data of 2020, it is stated that child workers have reached 160 million. It has been reported that bribery in developing countries is five times higher than in developed countries. It has been reported that from 2019 to 2020, 331 human rights defenders from 32 different countries were killed. It is stated that there are independent national human rights advocacy organizations established in accordance with international standards in only 82 countries around the world [57, 85].

The United Nations has implemented the sixteenth development plan to ensure peace among all the countries of the world, to conduct the judicial processes fairly and independently and to build strong institutions. Within the scope of the 16 Development Goal, the steps to be taken regarding justice, peace and strong institutional structures have been gathered under 12 headings:

- It should be aimed to prevent violence and all types of violence, which is one of the most important social problems, and to reduce deaths due to violence.

- It should be aimed to prevent international crimes such as child abuse and human trafficking, which are among the most important problems today, and to end violence and torture in all its dimensions.
- It should aim to operate the law mechanism on a global scale, make the rule of law work and equal access to the law for all societies. The rule of law must be ensured to ensure social justice.
- Due to the recent wars, the illegal circulation of valuable works and goods, especially the arms trade, has increased. By 2030, the focus should be placed on preventing them, returning stolen goods and artefacts, and combating all acts of organized crime.
- It should be aimed to combat and reduce bribery attempts and corruption at the national and international levels.
- The transparent management of institutions and organizations and the accountable structuring of these institutions should be encouraged.
- It should be aimed to increase participation in all institutions and organizations on a global scale, and to disseminate a comprehensive and representative management approach.
- Necessary steps should be taken so that developing countries can take part in the management processes of international organizations.
- At the international level, by 2030, all people should have a legal citizenship (identity) card, because, thanks to these identity cards, they can benefit from their citizenship rights and are recognized in all social institutions.
- In today's information societies, it should be aimed to protect everyone's access to information and basic citizenship and freedom rights within the scope of national and international laws. Access to information as a basic human right is a necessity for the sustainability of human life.
- Thanks to international partnerships, necessary steps should be taken to combat all kinds of crime and terrorist acts and groups, especially in developing countries, and the capacities of countries should be increased for this purpose.
- Comprehensive and non-discriminatory laws, regulations, plans and policies should be encouraged and used effectively for sustainable development.

Peace, justice and strong institutions are the three essential elements necessary to build a just and egalitarian society. These three elements are interconnected and form the basis of a thriving and stable community. Peace means the absence of violence, conflict and fear. When people live in peace, they can focus on their personal and professional goals without worrying about their safety and security. Peace also promotes social cohesion and trust, which is crucial for creating a sense of belonging and solidarity within a community.

Justice is about ensuring that everyone is treated fairly and equally. It includes promoting the rule of law, protecting human rights and promoting equality and inclusion. Justice also includes providing access to transparent, accountable and effective legal and judicial systems.

Strong institutions are those that can deliver basic public services and promote the rule of law. They are characterized by good governance, transparency and accountability. Strong institutions help build trust and confidence in government and its ability to serve the needs of its citizens. Together, these three principles form a strong framework for building a peaceful and just society. When people live in peace and security, they are more likely to achieve justice and benefit from strong institutions. When justice is served and institutions are strong, people are more likely to feel safe and protected. Achieving peace, justice and strong institutions requires the active participation of all members of society. It requires a commitment to democracy, human rights and the rule of law. It also requires investment in education, social services and infrastructure, as well as in the development of effective and accountable public institutions.

1.1.17 Partnership for the Goals (Strengthen the Means of Implementation and Revitalize the Global Partnership for Sustainable Development)

The seventeenth development goal gained recognition as a fundamental part of contributing to sustainable development through cross-sectoral partnerships. These partnerships focus on contributing to a sustainable society by collaborating through the knowledge, resources and expertise of businesses, institutions and organizations. All sectors have different knowledge and experience. Gathering these experiences around common goals is very important for sustainable development goals [11, 44, 88]. Therefore, the United Nations reveals the necessity of cooperation on a global scale in line with its 17th Development Goal. In this context, the steps to be taken on the topic of cooperation for purposes within the scope of the 17th Development Goal are gathered under 19 headings:

- For balanced and sustainable development, it should be aimed to use resources and create new resources, especially in developing countries. In addition, necessary steps should be taken to improve and develop taxation systems and other public collection systems.
- Balanced and sustainable development should be aimed at. Differences and imbalances in gross national product on a global scale must be eliminated. The level of economic and social welfare should be brought to international standards.
- For a balanced economic and financial distribution on a global scale, it should be aimed to increase financial opportunities, especially for developing countries.
- In particular, the borrowing processes of developing countries should be controlled. Necessary steps should be taken to repay debts and provide financial resources. Regulation and restructuring of existing debts should be encouraged, especially during periods of economic recession. In this way, the economic cycle can be managed in a healthy way. It is also necessary to reduce and control debt.

- Plans and policies should be established and incentives should be provided for the realization of global investments, starting with the least developed countries, for a fair and equal development process.
- Strengthening and access to information and technology infrastructures at regional, national and international levels on a global scale should be facilitated. In addition, a global technology infrastructure should be established, thanks to international cooperation and partnerships under the leadership of the United Nations.
- It should be aimed to build environmentally friendly technological infrastructures and to expand the use of these technologies on a global scale. For this, international agreements and conventions should be made and implemented.
- Opportunities should be created for developing countries in many subjects, especially scientific studies and technological infrastructures. In this way, a balanced development process can be built. In addition, until 2017, it should be aimed to disseminate the technology used in facilitating social processes.
- On a global scale, cooperation and partnerships should be developed in order to achieve all the goals set by the United Nations. For this, national plans should be created and capacity-building activities should be aimed.
- It should be aimed to create an open, transparent, comprehensive and non-discriminatory universal trade network that covers all societies. In this process, the principles of the World Trade Organization should be acted upon. As an indicator, the Doha Development Agenda items can be taken as a reference.
- Exports are the most important source of national income for countries. Therefore, it should be aimed to increase export opportunities starting from developing countries until 2020. In this way, production and sales processes are not collected in certain regions or countries, and global distribution is ensured.
- It should be aimed to consider restrictions and legal difficulties related to origin in import processes from less developed countries and to solve these problems. In addition, concrete steps should be taken to solve quota and customs problems. In this process, it is possible to act in accordance with the World Trade Organization Principles.
- Strategic steps can be taken by creating plans and policies regarding macroeconomic processes and indicators on a global scale. Consistent policies, especially at the international level, will be the most important indicator of this process.
- In order for the development processes to be sustainable, care should be taken to ensure that the policies are adequate and consistent.
- It can be aimed to create new and adequate policies to combat poverty, underdevelopment and imbalances in development. The existence of new practices and strategies in the developing and changing world is indisputable. However, in this process, due care and respect should be given to national policies and practices.
- It can be aimed to provide the necessary support and opportunities so that all societies and countries can reach their development goals and contribute. For this, information, technology and financial resources should be used, because global partnerships and collaborations seem possible for the development process to be sustainable.

- In addition to cooperation between countries, it should be aimed to establish and encourage institutional cooperation and partnerships regardless of the public or private sector.
- Regardless of the level of development, it should be aimed to obtain up-to-date, qualified data by 2020 in all countries and to reach these data in a timely manner. Different classification and discrimination (gender, religion, race, etc.) actions can be tracked and controlled, thanks to these updated data.
- A provincial statistical database should be created in order to follow the development of developing countries until 2030, and it should be aimed to define the steps taken for development.

Collaboration and partnerships are key to development on a global scale. This is extremely important for the 17th goal to become operational. It especially focuses on cooperation in mobilizing and increasing global capacity, ensuring financial and debt sustainability, and dissemination and transfer of technology [15].

Partnership for goals is a crucial aspect of achieving sustainable development in our world today. It includes building strong and inclusive partnerships between governments, non-governmental organizations, the private sector and other stakeholders to work together to achieve the 17 Sustainable Development Goals (SDGs) set by the United Nations. The SDGs cover a wide range of topics, including poverty, hunger, health, education, gender equality, clean water and sanitation, affordable and clean energy, decent work and economic growth, and climate action. Achieving these goals requires a comprehensive approach that addresses the underlying causes of these challenges and engages all sectors of society. Ultimately, partnership for goals is an essential component of achieving sustainable development. It involves building inclusive partnerships that leverage the shared power of all stakeholders to achieve the SDGs. With this collaborative approach, we can create a more sustainable and equitable world for all.

The development of countries will be possible as a result of complying with the roadmaps for these 17 goals set by the United Nations. The main purpose of the targets and indicators is to ensure that they catch up with the developing or developed countries, especially alongside underdeveloped countries. The re-emergence of Keynesian discourse after the 2008 crisis has led to the more widespread application of the SDGs than the MDGs [12].

While the world has been fluctuating during the intense globalization process since 1980, income inequalities have generally tended to increase with the emergence of the digital age. Since the development of each country is different, it turns out that a single structural change will not enable countries to achieve their goals. Since each country is different from the others, they have tried to develop accessible and feasible solutions by considering their own national and infrastructure problems, and have tried to increase the rate of achieving their goals by forming partnerships in line with the principle of inclusiveness in the context of the realization of universal goals [62].

In order for the SDGs to be sustainable and achievable, some criteria need to be taken into account. These are criteria such as planning measures, establishing monitoring mechanisms, evaluating progress, improving infrastructure and verifying data.

In order for the SDGs to be continuous and focused on development and improvement, it is necessary to go over the targets and their realization rates in a certain period of time.

All 17 goals are interconnected and play an important role in the development of societies. The SDGs go one step further towards ending poverty and target all nations and communities, developed and underdeveloped, rich or poor. These goals are more detailed in content. The SDGs aim to focus on underdeveloped or underdeveloped societies. They aim to strengthen better planned and organized initiatives by allowing stakeholders to enhance the needs and interests of those they represent.

1.2 Sub-dimensions of Sustainable Development

All societies on a global scale deserve economic and social development. It is a priority issue to ensure equal access to existing resources, especially while ensuring peace and tranquillity. In this process, a fair and balanced distribution will be decisive in solving the problems in societies. As a matter of fact, sustainable development is generally expressed as economic progress and development as well as the protection of nature and environmental quality [28]. At this point, it can be said that the development processes of countries and societies are a whole and consist of three sub-dimensions that support each other, because financial stability and development can also be achieved by considering social and environmental factors. For example, there is a planned use of natural resources so that production processes can turn into economic value. And a balanced reflection of production processes on societies can be considered as the key to social development.

The classical development approach's ignoring the environment in the name of economic development and bringing an individual economy-based understanding to the fore brought along some environmental problems. This understanding, which commodifies man's relationship with nature, necessitated the revision of the human-economy-based paradigm. Thus, underdevelopment, the industrialization process that damages natural resources, and the sustainable development approach that tries to balance with the environment and aims to ensure the justice of social welfare between present and future generations have been redesigned; defined by three dimensions as economy, environment and society. Considering all its dimensions, it is among the aims of international platforms to make sustainable development more usable at all levels, by making economic, social and environmental aspects interrelated and recognizing the connections between them, in order to achieve sustainable development [79].

Sustainability, the preservation and transfer of resources for future generations; in present day refers to development and progress, while also encompassing the sustainability of societies and the planet. Ultimately, encouraging the conscious and planned consumption of limited natural resources is one of the most important steps to be taken for a sustainable future. While meeting today's needs, ensuring the continuity of future generations without worrying about their existence and protecting

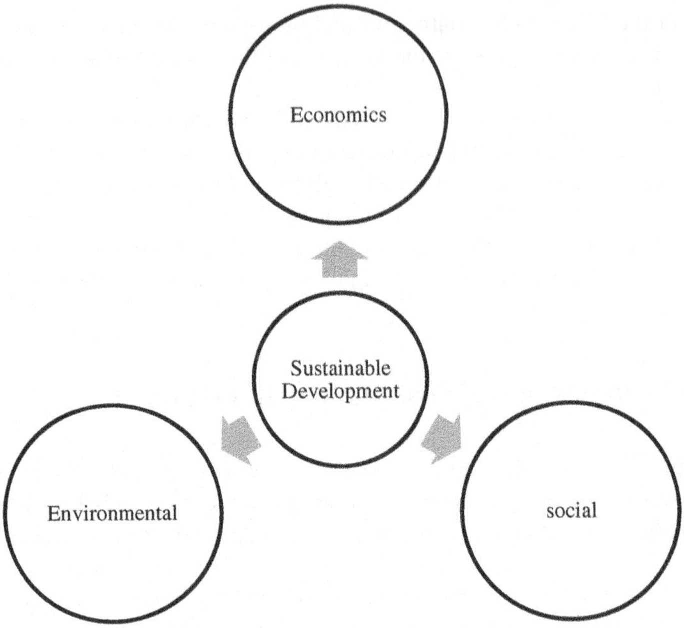

Fig. 1.2 Sub-dimensions of sustainable development. *Resource* Rogers et al. [61]

resources forms the basis of the principle of sustainable development. In this context, sustainable development consists of three sub-dimensions and these are economic, social and environmental sustainability dimensions. These three dimensions have a structure that is interconnected, influencing and supporting each other [61]. The three sub-dimensions of sustainable development are presented visually as follows (Fig. 1.2).

The environmental dimension draws attention to human activity, and the quality of human life (purity of the air we breathe, etc.) of the ecosystem operating in the region. The economic dimension, on the other hand, aims at increasing the efficient use of resources and increasing the welfare level of individuals in order to maximize market value. Finally, the social dimension focuses on the needs of the individual, the need for the majority, public awareness and harmony, and ultimately the social welfare of the people, with an emphasis on equality [80].

1.2.1 Economic Sustainability

Economic performance affects economic conditions as well as economic systems at the local, national or international levels. Especially the concept of financial performance and profit is vital for businesses. However, businesses also have to carry out some non-financial activities such as social activities and environmental measures [64]. The economic dimension, which expresses how the limited resources on earth will be distributed in the most equitable way in order to provide people with a better quality of life, and which has the capacity to create income and employment for the protection of the population, is a sub-dimension based on the protection of capital and preventing its deterioration [1, 39]. There is a direct relationship between sustainable development and economic development in terms of ensuring the welfare of society.

Sustainable development follows a moderate and encouraging strategy that aims to ensure that scarce resources in nature can be used by future generations. Economic growth, on the other hand, does not limit the use of resources, but focuses on the understanding of profit and efficiency [48]. The economic dimension of sustainable development is accepted as keeping natural resources and capital intact [25]. In short, it has been seen that integrating environmental issues into the enterprise strategy of enterprises provides a number of benefits that reduce costs and improve the efficiency of the enterprise, improve product quality, increase market share, put it ahead of laws and competitors, create new markets, increase employee motivation and satisfaction, and also provide financial aid [93].

At the same time, it is a system that constantly produces goods and services, avoids sectors that harm industrial and agricultural production and ensures the sustainability of domestic and foreign debt at manageable levels economically [21]. It also requires the management of various economic capitals of the business such as financial, tangible and intangible capital. Economically sustainable firms guarantee a permanent cash flow while providing adequate support to their stakeholders [22].

There are four basic features that define the economic dimension of sustainable development. These four basic features determine the explanatory qualities of the economic dimension of sustainable development. These key features are as follows:

- Orientation towards a long-term future in which a great deal of uncertainty remains.
- Considering nature and human relations while creating a work.
- Establishing a just relationship between nature and man for present and future generations.
- It is the understanding of the concerns for economic activity as a waste of time in the allocation of natural goods and service substitutes [8].

In this context, economically sustainable development means the continuous improvement of human welfare, the increase of capital or at least the protection of it. Therefore, the effects of limited scarce resources on economic growth, the effective distribution of these resources, technological developments and the substitution of existing resources require evaluation [92]. As a matter of fact, sustainable development represents an integral part of sustainability in terms of environmental and social aspects. This situation also clearly reveals the limits of economic growth [58]. Therefore, businesses should also have social and environmental sustainable policies. However, it is not possible to ensure sustainability without the economic activities of enterprises, which are the building blocks of economic life [40].

1.2.2 Social Sustainability

The interpretations and statements about the social dimension of sustainable development are as diverse as the comments and statements about economic sustainability. Social sustainability can be defined as the extent to which social values, social identities, social relations and social institutions can continue into the future. The focus of the social dimension of sustainable development is on the development and well-being of the society, its constructions and the groups of people [26]. The social dimension of sustainability refers to the spread of unity and integrity throughout society and to act together and unite around common interests. According to the scope of social sustainability, cultural and educational activities refer to adequately meeting the needs of health and social well-being and housing. Social development and sustainability are extremely important in the creation of collective consciousness [49]. It is inevitable to ensure equality for social groups, which have an important effect on the continuity of social and cultural systems, and to create an environmental organization system that alleviates poverty [66]. Social sustainability includes processes that speak for the creation and conditions of positive conditions for all societies. This definition is comprehensive but includes the following:

- Fair and equitable access to services
- Eliminating generational differences and ensuring equality
- Supporting and valuing diversity in society
- Wide participation in politics and policy processes
- Collective consciousness
- Creating an awareness network for the development and dissemination of the understanding of social sustainability
- Building a self-sufficient society
- Political attitude and advocacy in meeting social needs [50].

The social sustainability dimension focuses on the needs of the individual, the need for the majority, public awareness and harmony, as a result, the social welfare of the people, as stipulated by the United Nations, with the emphasis on equality [81]. A pluralistic social structure is possible with information accessibility, institutional reputation and stability, and the protection of cultural identity and socio-cultural wealth [33]. It is essential to build a polyphonic structure rather than monophony in society and to prepare a quality living ground [45]. And it is the mainstay of the social sustainability dimension.

Social sustainability aims to enrich individual social welfare. The participatory and pluralistic social structure can be accepted as a ground where differences are transformed into wealth. However, a transparent, fair and accessible level of social welfare can also be seen as a step in the development of humanity. Equal opportunities, and equal access to resources and development tools can facilitate the minimization of differences in society. In this way, the differences can be evaluated as an element of wealth without sharpening.

1.2.3 Environmental Sustainability

Environmental sustainability can be defined as creating and maintaining conditions in which humans and nature can exist in productive harmony and that allow the social, economic and environmental needs of future generations to be met [75], and also to meet human needs without compromising the health of ecosystems [51].

Environmental sustainability has been a hot topic recently. It can be said that it is an area of theory and practice that has become the problem of all societies, especially climate change and global warming. The environmental costs of human life, production and consumption activities are growing. It can be stated that the unplanned and irregular use of existing natural resources for production processes constitutes the basis of this problem. The perception and use of natural resources as if they are unlimited have begun to produce devastating results. All vital activities such as increased human mobility, intensification of production and the growth of the consumption chain associated with it have become a threat to the existence of the planet we live in. Bansal [7] draws attention to the fact that if the natural environment we live in is at risk, the vital resources such as air, water and food that support the existence of human beings will also be seriously threatened. The understanding of environmental sustainability is the most appropriate tool to bring sustainable development together with future generations. However, since human activities impair the ability of the environment to renew itself, this situation not only prevents the welfare of future generations but also threatens their right to exist [73].

Environmental sustainability is an integral part of sustainable development. The environment is a unique stakeholder in economic and social systems, as it is at the centre of all human activities. However, the activities of all social structures negatively affect the environment as stated above [56]. However, in recent years, especially commercial enterprises have realized that the relationship between them and the environment is not very healthy. In this context, necessary steps have been taken and the logic of green and clean management has started to become widespread, because the implementation of environmentally friendly policies increases operational efficiency and thus serves as a superior source of competitive advantage [17]. It is not possible for society to meet its needs without the environment and environmental values. Therefore, the most important dimension in sustainable development is the environment. The environmental dimension in sustainable development includes increasing the values beyond protecting the environment and the natural resources in the environment [54]. In this context, the applications that businesses have been trying to put forward in the last period are not just commercial. In addition, plans, policies and practices that bring environmental concerns to the fore are expressed loudly. The objectives related to environmental sustainability and the basic principles of environmental sustainability can be listed as follows:

- Controlling existing natural resources and reducing impacts on the ecosystem.
- Establishing knowledge-based systems in decision-making processes.
- Ensuring communication between the environment and the social system.
- Strengthening the management structure for global environmental targets.

- Dissemination and promotion of recycling.
- To prevent the release of threatening harmful wastes, especially chemicals, into nature.
- To act in a controlled manner in the consumption of non-recyclable and non-renewable resources.
- Creating long-term plans and policies.
- Establishing and operating a feedback system on environment and sustainability.
- Using different assessment and scale systems.
- Demonstrate a flexible understanding in solving problems.
- Respecting the environment and nature in all its aspects [49].

As the subject of this research, environmental sustainability is the most important resource of the planet and human life. There are not only economic concerns in the juxtaposition of sustainability and the environment. Today, all sectors create negative environmental effects due to production activities. Therefore, both an unplanned consumption of natural resources and an unlimited production cycle occur. This cycle may make human life easier, but it has devastating effects on the environment and nature. One of the sectors where these effects have become increasingly evident in the recent period is the global sports industry. Sports is one of the sectors where intense human mobility is seen. A continuous production cycle, endless leagues and organizations, and national and international races create huge human mobility around the world. People constantly travel for sports events and competitions, feed waste generation and create environments for their energy use. Therefore, such massive mobility leaves significant effects on the environment and nature. In the next section, a comprehensive assessment of the negative environmental impacts of the sports industry and its effects is presented.

References

1. E. Aba Şenbayram, Sürdürülebilir kalkınma perspektifinden yeşil finans. Sosyal Bilimler Araştırma Dergisi **11**(3), 399–409 (2022)
2. A. Akgün, Ü. Duruk, F. Tokur, Gözlem gezisi yönteminin öğretmen adaylarının çevreye ve çevre eğitimine ilişkin görüşlerine etkisi. Route Educ. Soc. Sci. J. **4**(2), 65–82 (2017)
3. B. Anger, Poverty eradication, millennium development goals and sustainable development in Nigeria. J. Sustain. Dev. **3**, 138–144(2010). https://hdl.handle.net/10535/6830
4. B. Aquilani, C. Silvestri, G. Ioppolo, A. Ruggieri, The challenging transition to bio-economies: Towards a new framework integrating corporate sustainability and value co-creation. J. Clean. Prod. **172**, 4001–4009 (2018)
5. A.D. Atik, Y. Doğan, Lise öğrencilerinin küresel iklim değişikliği hakkındaki görüşleri. Acad. J. Educ. Sci. **3**(1), 84–100 (2019)
6. F. Aydın, Ortaöğretim öğrencilerinin küresel ısınma konusundaki bilgi düzeylerinin belirlenmesi. Turkish J. Educ. **3**(4), 15–27 (2014)
7. P. Bansal, Evolving sustainably: a longitudinal study of corporate sustainable development. Strateg. Manag. J. **26**(3), 197–218 (2005)
8. M.Y. Bilgili, Ekonomik, ekolojik ve sosyal boyutlarıyla sürdürülebilir kalkınma. Uluslararası Sosyal Araştırma Dergisi **10**(49), 559–569 (2017)

9. H.P. Blossfeld, E. Kilpi-Jakonen, D. Vono de Vilhena, S. Buchholz, *Adult Learning in Modern Societies: an International Comparison from a Life-Course Perspective* (Edward Elgar, 2014)
10. H. Borland, Y. Bhatti, A. Lindgreen, Sustainability and sustainable development strategies in the U.K. plastic electronics industry. Corp. Soc. Responsibil. Environ. Manag **26**(4), 805–818 (2019). https://doi.org/10.1002/csr.1722
11. K. Buhmann, J. Jonsson, M. Fisker, Do no harm and do more good too: connecting the SDGs with business and human rights and political CSR theory. Corporate Governance. Int. J. Bus. Soc. **19**(3), 389–403 (2019)
12. J.B. Carant, Unheard voices: a critical discourse analysis of the millennium development goals' evolution into the sustainable development goals. Third World Q **38**(1), 16–41 (2017)
13. F.S. Chaphin, M. Torn, M. Tateno, Principles of ecosystem sustainability. Am. Nat. **6**(148), 1016–1037 (1996)
14. T.M. Cojocaru, G.H. Ionescu, D. Firoiu, L.M. Cismaş, M.D. Oţil, O. Toma, Reducing inequalities within and among EU Countries—Assessing the achievement of the 2030 Agenda for sustainable development targets (SDG 10). Sustainability **14**(13), 7706 (2022). https://doi.org/10.3390/su14137706
15. N. Cooper, D. French, SDG 17: partnerships for the goals-cooperation within the context of a voluntarist framework, in ed. by D. French, L.J. Kotzé *Sustainable Development Goals* (1st edn. Edward Elgar Publishing, Cheltenham, 2018), pp. 271–304
16. R. Damania, S. Desbureaux, A.S. Rodella, J. Russ, E. Zaveri, *Quality Unknown: the Invisible Water Crisis* (2019). https://openknowledge.worldbank.org/bitstream/handle/10986/32245/9781464814594.pdf?sequence=8&isAllowed=y. Accessed 9 Mar 2023
17. A. Danso, S. Adomako, T. Lartey, J. Amoah, D. Yirenkyi, Stakeholder ıntegration, environmental sustainability orientation and financial performance. J. Bus. Res. **119**, 652–662 (2020). https://doi.org/10.1016/j.jbusres.2019.02.038
18. M. Demircan, N. Turan, H. Arabacı, M. Coşkun, N. Türkoğlu, İ. Çiçek, Gündem belirleme modeline göre yazılı basındaki iklim değişikliği haber ve köşe yazılarının analizi. TÜCAUM Uluslararası Coğrafya Sempozyumu, 13–14 Ekim 2016 (Ankara, 2016)
19. J. Denoncourt, Companies and UN 2030 sustainable development goal 9 industry, innovation and infrastructure. J. Corp. Law Stud. **201**(1), 199–235 (2020). https://doi.org/10.1080/14735970.2019.1652027
20. S. Dhahri, A. Omri, Foreign capital towards SDGs 1 & 2—Ending poverty and hunger: the role of agricultural production. Struct. Chang. Econ. Dyn. **53**, 208–221 (2020). https://doi.org/10.1016/j.strueco.2020.02.004
21. M. Doğruel Anuşlu, S.Ü. Fırat, Ülkelerin endüstri 4.0 seviyesinin sürdürülebilir kalkınma düzeylerine etkisinin analizi. Endüstri Mühendisliği Dergisi **31**(0), 44–58 (2020)
22. T. Dyllick, K. Hockerts, Beyond the business case for corporate sustainability. Bus. Strateg. Environ. **11**(2), 130–141 (2002). https://doi.org/10.1002/bse.323
23. M.V. Eren, Cinsiyet eşit(siz)liğinde genç kadın işsizliği ile kalkınma arasındaki ilişki: Avrupa Birliği ülkeleri üzerine ekonometrik bir analiz. Yaşar Üniversitesi E-Dergisi **15**(59), 598–614 (2020). https://doi.org/10.19168/jyasar.702505
24. G. Giuliani, B. Chatenoux, A. Benvenuti, P. Lacroix, M. Santoro, P. Mazzetti, Monitoring land degradation at national level using satellite earth observation time-series data to support SDG15-exploring the potential of data cube. Big Earth Data **4**(1), 3–22 (2020)
25. R. Goodland, W. Bank, Sustainability: human, social, economic and environmental. ted munn. Encycl. Glob. Environ. Change **6**, 220–225 (2002)
26. S. Gürlük, Sürdürülebilir kalkınma gelişmekte olan ülkelerde uygulanabilir mi? Eskişehir Osmangazi Üniversitesi İİBF Dergisi **5**(2), 85–99 (2010)
27. N.L. Hall, S. Creamer, W. Anders, A. Slatyer, P.S. Hill, Water and health interlinkages of the sustainable development goals in remote indigenous Australia. Clean Water **3**, 1–10 (2020). https://doi.org/10.1038/s41545-020-0060-z
28. J.M. Harris, B. Roach, Environmental and natural resource economics: a contemporary approach. ME Sharpe

29. R. Hillerbrand, Why affordable clean energy ıs not enough. A capability perspective on the sustainable development goals. Sustainability **10**(7), 2485 (2018). https://doi.org/10.3390/su10072485
30. Z. Hossen, B.H. Khondker, How is India flaring in achieving SDG 10 on reduced inequality? In *Sustainable Development Goals* (Springer, 2020)
31. IEG World Bank, Consultative Group to Assist the Poor (2008). https://ieg.worldbankgroup.org/sites/default/files/Data/reports/gpr_cgap.pdf. Accessed 5 Mar 2023
32. International Council for Science, A Guide to SDG Interactions: from Science to Implementation. (International Council for Science (ICSU), Paris, 2017)
33. M.A. Kahn, Sustainable development: the key concepts, issues and implications. Sustain. Dev. **3**, 63–69 (1995)
34. R. Karalar, H. Kiracı, Çevresel sorunlara karşı bir çözüm önerisi olarak sürdürülebilir tüketim düşüncesi. Dumlupınar Üniversitesi Sosyal Bilimler Dergisi **30**, 63–76 (2015)
35. Ş Kaypak, Küreselleşme sürecinde sürdürülebilir bir kalkınma için sürdürülebilir bir çevre. KMÜ Sosyal ve Ekonomik Araştırmalar Dergisi **13**(20), 19–33 (2011)
36. S. Khetrapal, R. Bhatia, Impact of COVID-19 pandemic on health system & sustainable development goal 3. Indian J. Med. Res. **151**(5), 395–399. https://doi.org/10.4103/ijmr.IJMR_1920_20
37. M.Z. Koç, Ç. Sel, Sürdürülebilir üretim çizelgeleme problemlerinde insan faktörü. Int. J. Eng. Res. Dev. Special Issue, 130–136 (2021). https://doi.org/10.29137/umagd.1014876
38. N. Kong, O. Salzmann, U. Steger, A. Ionescu-Somers, Moving business/industry towards sustainable consumption: the role of NGOs. Eur. Manag. J. **20**(2), 109–127 (2022)
39. Ç. Koşar Taş, S. Örk Özel, D. Veysikarani, Dijital ekonominin sürdürülebilir kalkınma hedefleri üzerine etkisi: Avrupa Birliği ülkeleri için bir inceleme. Abant Sosyal Bilimler Dergisi **21**(2), 55–78 (2021)
40. N. Kuşat, Sürdürülebilir işletmeler için kurumsal sürdürülebilirlik ve içsel unsurları. Afyon Kocatepe Üniversitesi İİBF Dergisi **14**(2), 227–242 (2012)
41. I. Lapinskaitė, S. Vidžiūnaitė, Assessment of the sustainable economic development goal 8: decent work and economic growth in G20 Countries. Econ. Cult. **17**(1), 116–125 (2020). https://doi.org/10.2478/jec-2020-0011
42. W. Leal Filho, L.L. Brandli, A. Lange Salvia, L. Rayman-Bacchus, J. Platje, COVID-19 and the UN sustainable development goals: threat to solidarity or an opportunity? Sustainability **12**(13), 1–14 (2020). https://doi.org/10.3390/su12135343
43. N. Lövehagen, A. Bondesson, Evaluating sustainability of using ICT solutions in smart cities–methodology requirements, in *International Conference on Information and Communication Technologies for Sustainability*, pp. 175–182
44. A. MacDonald, L. Clarke, M. Huang, M. Roseland, M.M. Seitanidi, Multi-stakeholder partnerships (SDG #17) as a means of achieving sustainable communities and cities (SDG #11), in ed. by W. Leal Filho Handbook of sustainability science and research, 1st edn. (Springer International Publishing, 2018), pp. 193–209
45. S. McKenzie, Social sustainability: towards some definitions, in *Hawke Research Institute, Working Paper Series* (Vol. 27, 2004), pp. 1–31
46. W.W. McMahon, The external benefits of education, in eds. by P. Peterson, E. Baker, B. McGaw *International Encyclopaedia of Education*, 3rd edn. (Elsevier, 2010), pp. 260–271)
47. D.H. Meadows, D.L. Meadows, F. Randers, W.W. Behrens, The limits to growth: a report for the club of Rome's project on the predicament of mankind. Int. Kindle Paperwhite (1974)
48. K. Mishra, A. Bais, Sustainable development: roadmap for sustainable future. UGC Care Group I J. **10**(7), 76–80 (2020)
49. B. Moldan, S. Janousková, T. Hak, How to understand and measure environmental sustainability: indicators and targets. Ecol. Ind. **17**, 4–13 (2012)
50. J. Morelli, Environmental sustainability: a definition for environmental professionals. J. Environ. Sustain. **1**(1), 1–10 (2011)
51. L.F. Mortensen, Sustainable household consumption in Europe? Consum. Policy Rev. **16**(4), 4–17 (2006)

52. D. Newhouse, P. Suarez-Becerra, M. Evans, New estimates of extreme poverty for children, in *Policy Research Working Paper* (Vol. 7845, World Bank: Washington, DC, 2016)
53. K.E. Özkan, Sürdürülebilir kalkınma bağlamında çevre sorunlarının önemi: Türkiye ve AB karşılaştırması (Master's thesis, Bilecik Şeyh Edebali Üniversitesi, Sosyal Bilimler Enstitüsü, 2017)
54. V. Paloma, M. Escobar-Ballesta, B. Galván-Vega, J.D. Díaz-Bautista, I. Benítez, Determinants of life satisfaction of economic migrants coming from developing countries to countries with very high human development: a systematic review. Appl. Res. Qual. Life **16**, 435–455 (2021)
55. M. Prasad, T. Mishra, V. Bapat, Corporate social responsibility and environmental sustainability: evidence from India Using energy intensity as an indicator of environmental sustainability. IIMB Manag. Rev. **31**(4), 374–384 (2019)
56. S. Rassanjani, Ending poverty: factors that might influence the achievement of sustainable development goals (SDGs) in Indonesia. J. Public Adm. Gov. **8**(3), 114–128 (2018). https://doi.org/10.5296/jpag.v8i3.13504
57. Rebuplic of Turkey Presidency of Strategy and Budget, Sürdürülebilir kalkınma amaçları değerlendirme raporu (2019). https://www.sbb.gov.tr/wp-content/uploads/2020/03/Surdurulebilir-KalkinmaAmaclariDegerlendirme-Raporu_13_12_2019-WEB.pdf. Accessed 13 Mar. 2023
58. T.L. Reddy, R. J. Thomson, Environmental, social and economic sustainability: implications for actuarial science. Actuar. Inst. 23–27 (2015)
59. Republic of Turkey Ministry of Foreign Affairs, Birleşmiş Milletler İklim Değişikliği Çerçeve Sözleşmesi (2010). http://www.mfa.gov.tr/bm-iklim-degisikligi-cerceve-sozlesmesi.tr. Accessed 4 Mar. 2023
60. W. Rickels, J. Dovern, J. Hoffmann, M.F. Quaas, J.O. Schmidt, M. Visbeck, Indicators for monitoring sustainable development goals: an application to oceanic development in the European Union. Earth's Fut. **4**(5), 252–267 (2016). https://doi.org/10.1002/2016EF000353
61. P.P. Rogers, K.F. Jalal, J.A. Boyd, An introduction to sustainability development. Glen Educ. Found. (2008)
62. J.D. Sachs, From millennium development goals to sustainable development goals. Lancet, 2206–2211 (2012)
63. T. Schuller, J. Preston, C. Hammond, A. Brassett-Grundy, J. Bynner, *The benefits of learning: the impact of education on health, family life and social capital* (Routledge, 2004)
64. B. Sezen, Y.S. Çankaya, Effects of green manufacturing and eco-innovation on sustainability performance. Procedia Soc. Behav. Sci. **99**, 154–163 (2013)
65. A. Sodiq, A.B.B. Ahmer, A.K. Shoukat, N. Sezer, M. Seif, J. Mohamoud, A. Ali, Towards modern sustainable cities: review of sustainability principles and trends. J. Clean. Prod. **277**, 972–1001 (2019)
66. J.E. Stiglitz, Employment, social justice and societal well-being. Int. Labour Rev. **141**(1–2), 9–29 (2002)
67. A. Stratigea, C.A. Papadopoulou, M. Panagiotopoulou, Tools and technologies for planning the development of smart cities. J. Urban Technol. **22**(2), 43–62 (2015)
68. H. Suzuki, A. Dastur, S. Moffatt, N. Yabuki, H. Maruyama, in *Eco2 Cities: Ecological Cities as Economic Cities* (World Bank, 2010)
69. W.M. Sweileh, Bibliometric analysis of scientific publications on "sustainable development goals" with emphasis on "good health and well-being" goal (2015–2019). Glob. Health **16**(68), 1–13 (2020). https://doi.org/10.1186/s12992-020-00602-2
70. J.K. Swim, S. Clayton, G.S. Howard, Human behavioural contributions to climate change psychological and contextual drivers. Am. Psychol. **66**(4), 251–264 (2011)
71. K. Szopik-Depczyńska, A. Kędzierska-Szczepaniak, K. Szczepaniak, K. Cheba, W. Gajda, G. Ioppolo, Innovation in sustainable development: an investigation of the EU context using 2030 agenda indicators. Land Use Policy **79**, 251–262 (2018)
72. H. Tıraş, Sürdürülebilir kalkınma ve çevre: Teorik bir inceleme. Kahramanmaraş Sütçü İmam Üniversitesi İktisadi ve İdari Bilimler Fakültesi Dergisi, 57–73 (2012)

73. C. Tortajada, Contributions of recycled wastewater to clean water and sanitation sustainable development goals. Clean Water **3**, 1–22 (2020). https://doi.org/10.1038/s41545-020-0069-3
74. C. Tortajada, A.K. Biswas, Achieving universal access to clean water and sanitation in an era of water scarcity: strengthening contributions from academia. Curr. Opin. Environ. Sustain. **34**, 21–25 (2018). https://doi.org/10.1016/j.cosust.2018.08.001
75. U.S. Department Of Energy, Environmental sustainability (2020). https://www.energy.gov/lm/listings/environmental-sustainability. Accessed 13 Mar. 2023
76. UNCDF, *News Release: Dakar Conference on Financial Access for the Poor Opens to Widespread Optimism and Support* (United Nations Capital Development Fund, 2006)
77. UNICEF, Millennium Development Goals Monitoring (2015). http://www.un.org/millenniumgoals/2015_MDG_Report/pdf/MDG%202015%20rev%20(July%201).pdf. Accessed 5 Mar. 2023
78. United Nation ESCAP, Statistical Yearbook for Asia and the Pacific (2015). https://repository.unescap.org/bitstream/handle/20.500.12870/4175/ESCAP-2016-FS-SYB2015-SDG8.pdf?sequence=1&isAllowed=y. Accessed 9 Mar. 2023
79. United Nations, The future we want, in *Rio de Janerio: Rio +20 United Nations Conference on Sustainable Development* (2012). https://sustainabledevelopment.un.org/futurewewant.html. Accessed 13 Mar. 2023
80. United Nations, Transforming our world: the 2030 Agenda for Sustainable Development (2015). https://www.un.org/ga/search/view_doc.asp?symbol=A/RES/70/1&Lang=E. Accessed 23 Mar. 2023
81. United Nations, SDG-15. Life on land: why it matters (United Nations Sustainable Development Goals, New York, NY, 2016)
82. United Nations, Oceans: facts and figures (2017). http://www.un.org/sustainabledevelopment/oceans/. Accessed 13 Mar. 2023
83. United Nations Children's Fund (UNICEF), Water, sanitation and hygiene (WASH): annual (2013). https://www.unicef.org/wash/files/WASH_Annual_Report_Final_7_2_Low_Res.pdf. Accessed 9 Mar. 2023
84. United Nations Department of Economic and Social Affairs, Sustainable development goals (2015). https://sdgs.un.org/goals. Accessed 4 Mar. 2023
85. United Nations-The World Commission on Environment and Development, Our common future-brundland report. Report No.: A/42/427 (1987). https://www.are.admin.ch/are/en/home/media/publications/sustainabledevelopment/brundtland-report.html. Assessed 7 Mar. 2023
86. United Nations, Department of Economic and Social Affairs, Sustainable Development Outlook 2020 (2020). https://sdgs.un.org/sites/default/files/2020-07/SDO2020_Book.pdf. Assessed 4 Mar. 2023
87. United Nations, Copenhagen Declaration on Social Development: World Summit for Social Development (1995). https://www.un.org/development/desa/dspd/world-summit-for-social-development-1995.html. Assessed 4 Mar. 2023
88. I. Van Hille, F.G.A. De Bakker, J.E. Ferguson, P. Groenewegen, Cross-sector partnerships for sustainability: how mission-driven conveners drive change in national coffee platforms. Sustainability **12**(7), 2846 (2020)
89. World Bank, Improving on-site sanitation and connections to sewers in Southeast Asia insights from Indonesia and Vietnam. (2015). http://wsp.org/sites/wsp.org/files/publications/WSP-Improving-On-site-Sanitvation-Connectionsto-Sewers-Southeast-Asia.pdf. Assessed 9 Mar. 2023
90. World Bank, Rethinking infrastructure for development (2007). https://documents1.worldbank.org/curated/en/538461468175472629/pdf/414460ABCDE020101OFFICIAL0USE0ONLY1.pdf. Assessed 5 Mar. 2023
91. A. Yavuz, Sürdürülebilirlik Kavramı ve işletmeler açısından sürdürülebilir üretim stratejileri. Mustafa Kemal Üniversitesi Sosyal Bilimler Enstitüsü Dergisi **7**(14), 63–86 (2010)

92. O. Yeni, Sürdürülebilirlik ve sürdürülebilir kalkınma: Bir yazın taraması. Gazi Üniversitesi İktisadi ve İdari Bilimler Fakültesi Dergisi, 181–208 (2014)
93. Q. Zhu, J. Sarkis, Relationships between operational practices and performance among early adopters of green supply chain management practices in Chinese manufacturing enterprises. J. Oper. Manag. (2004). https://doi.org/10.1016/j.jom.2004.01.005

Chapter 2
Environmental Sustainability in Sport

The effort to accelerate social life and to ensure the balance of production and consumption and the widespread use of mass media brought a significant improvement in human life. However, this acceleration has increased the need for resources. In meeting this need, it may cause people to turn to natural resources and the environment more. In today's modern world, basic problems such as increasing population, migration towards cities due to employment opportunities, increase in production and consumption processes and the increase in natural resource consumption are waiting to be solved because these problems threaten the sustainable environment and human life [61]. Of course, the basis of this process is people and their attitudes and behaviours because the behaviours exhibited by people in their basic life cycles cause destruction on the environment [5].

The planet we live on has faced a climate change on a global scale. Sustainability of the environment is an issue that is discussed loudly among scientists, politicians and all other social structure representatives. Negative environmental effects caused by vital activities are handled through many different indicators with the understanding of life cycle assessment [26]. Basically, with the increase in human activities, emissions reach high levels and this situation forms the basis for climate change and global warming. The foremost of these environmental problems is the gradual warming of climates on a global scale due to high warming degrees. The rapid warming of climates on a global scale threatens ecosystems, biodiversity and natural vegetation [65].

In the modern world age, all human-based vital activities have increased. This increase has become the source of an important environmental problem. This problem is greenhouse gas and carbon emissions. Greenhouse gas and carbon emissions are the most important components of the global warming and climate change crisis [88]. Since natural resources do not rely on the increasing human population, they are the main source of many problems between man and the environment. Social problems, excessive consumption due to increasing human needs, poverty, pollution, loss of biodiversity, spread of new diseases and depletion of basic resources are seen

A. Atalay and B. Švagždienė, *Achieving Environmental Sustainability in the Sports Sector*, SpringerBriefs in Applied Sciences and Technology,
https://doi.org/10.1007/978-3-031-37401-2_2

as strategic threats to human civilization [21]. One of the indicators of sustainable development and atmospheric pollution is the carbon footprint [68].

The problem of global warming is increasing every day, so it is necessary to reduce greenhouse gas (GHG) emissions worldwide and avoid irreversible damages [76]. The areas used as settlement areas in the world constitute only 2% of the land surface. However, resource consumption is quite high even in this small area due to the dense population [2, 87]. In addition, cities are identified with environmental sustainability problems expressed in terms of traffic congestion, noise, air quality and GHG emissions. Therefore, up to the point reached, the plans, policies and strategies aim to reduce negative environmental impacts and reduce emissions due to greenhouse gas and carbon emissions. Alternatives have been tried to be developed for the solution of climate change and crisis [51].

All stakeholders of sports on a global scale are trying to be more careful about the environmental effects caused by sports. In this context, they have started to add the environmental title to their management policies in order to balance the relationship between sports and the environment and to minimize the negative effects [69, 90, 91] because measuring and reporting greenhouse gas emissions has become very important for the implementation of international policies/strategies to reduce and adapt to climate change [12, 49]. For example, the Rio and Kyoto Summits convened to take international steps for environmental sustainability resulted in national, regional and local governments and affiliates expressing their commitment to act in more environmentally sustainable ways [66]. Long before these global meetings and evaluation talks, Shrivastava introduced the idea of "ecocentric" management in 1995. After the emergence of this idea, its use has become widespread in many businesses, organizations and institutions. Businesses that have adopted the ecocentric management idea have started to establish a link between the natural environment in which they operate and their social environment and have aimed to act responsibly. On the basis of this management approach, it is aimed to minimize waste generation and pollution of the natural environment with the planned and regular use of natural resources [73].

The "ecocentric" management approach is the environmentally friendly management approach. Institutions and organizations, regardless of the areas in which they operate, act sensitively to environmental problems and aim to minimize their negative environmental impacts. For this reason, all stakeholders in all sectors (production, transportation, tourism, sports, etc.) create and try to implement plans and policies related to environmental sustainability in almost all management processes. This development has emerged entirely due to the negative development of the relationship between human and environment.

The sports industry is one of the sectors where people are at the centre and where intense human activities are seen. All sports events organized for health, tourism and sportive purposes attract great attention because they are human-based activities. In all sports branches, especially football, teams and fans participate in both national and international events and benefit from sports and products and services. For example, it is known that thousands or even millions of people participate in these competitions and events in popular sports branches. For example, according to FIFA data, while there are 265 million active football players all over the world, there are 4 billion active football fans, which corresponds to half of the world's population [78]. Therefore, there is mobility and participation of millions of people. As a matter of fact, considering that two matches are played every week in local football leagues, the intensity of this human activity is incredible.

The understanding of environmental sustainability in the sports sector is a concept that refers to the reduction of the negative environmental effects of sports. This includes reducing carbon emissions from transport and energy use, as well as reducing waste and water use. One way sports organizations and event organizers work for environmental sustainability is by taking sustainable measures in the construction, maintenance, repair and operation of sports facilities. Another point that sports clubs and federations can implement in environmental policies is to promote sustainable behaviours among fans and athletes. This may include promoting ridesharing or public transport to events, promoting sustainable products and services and raising awareness of environmental issues.

2.1 Environmental Issues in Sport Sector

The race of many countries to host organizations has gained momentum and has become important due to the global influence that the sports sector has. However, this increasing demand has brought along environmental concerns and negative environmental effects, and naturally, the sports sector has become an important part of the climate change problem [53]. For this reason, sports is one of the sectors where measures should be taken against threats related to environmental sustainability. As a matter of fact, the intensification of carbon and greenhouse gas emissions at the beginning of these threats, in other words, means the growth of the carbon footprint. Today, one of the most important sources of carbon and greenhouse gas emissions is the sports sector [63]. Sport is the catalyst of human behaviour that has negative effects on all segments of society and the environment due to its characteristics [54]. Regarding these negative effects of the sports industry, the United Nations reports that the carbon footprint in the sports industry is growing gradually due to the production of sports events, sports facilities, sports activities and sports equipment [85].

With the strengthening of the relationship between the sports industry and climate change and environmental problems on a global scale, academic literature has begun to emerge and research on this subject has diversified. Depending on this situation, especially the International Olympic Committee has started to take important steps. In order to address the growing concerns, "IOC Sustainability Strategy" [45]. Together with the International Olympic Committee, member countries take the necessary steps to eliminate environmental concerns and take initiatives to eliminate environmental problems in sports. One of these countries is Italy. Trying to accompany the steps of the International Olympic Committee in theory and practice, the "Italian Sports Association for All" has been conducting comprehensive reporting studies under the name of "sustainability in sports" since 2012. One of the main objectives of this reporting process is to determine how the sports industry poses a threat to environmental sustainability. The Association made a comprehensive assessment of environmental sustainability in the sports sector and gathered the sources of environmental risk in sports under two main headings. These are the establishment and organized organizations in sports. The evaluation classification is presented in the image below.

In the report prepared by the Association of Sports for All in Italy, where extensive studies on sports and environmental effects are carried out, the relationship between sports and environmental sustainability is grouped under two headings. The first of these is the establishment activities carried out to ensure the continuity of the sports sector. The other is the events, tournaments and activities organized at the national and international level. Depending on the facility, air, water and soil pollution as well as the wastes generated by the building processes come to the fore. In addition, the energy used in these facilities has been evaluated as a risk factor for the environment due to the carbon emissions it causes. The environmental effects of organized events and tournaments are considered as waste generation due to travel, accommodation and food consumption, and energy consumption [75] (Fig. 2.1).

There is a striking example of embodying and quantifying the negative environmental effects caused by the sports industry. This example is the World Cup organization held in Brazil. The World Cup held in Brazil produced about 2.7 million tonnes of greenhouse gas emissions, which is considered disproportionately high for an event that starts and ends in a month's time [32]. The carbon footprint of the 2014 World Cup held in Brazil is due to transportation (83.7%), stadiums and facilities (9.65%), tourist accommodation (5.7%) and various tourist activities (0.95%). The most important source of the carbon footprint that emerged in the World Cup has been transportation. Most of this carbon footprint comes from international (60.5%), long-distance domestic (35.2%) and short-distance domestic/local travel (4.3%) [32].

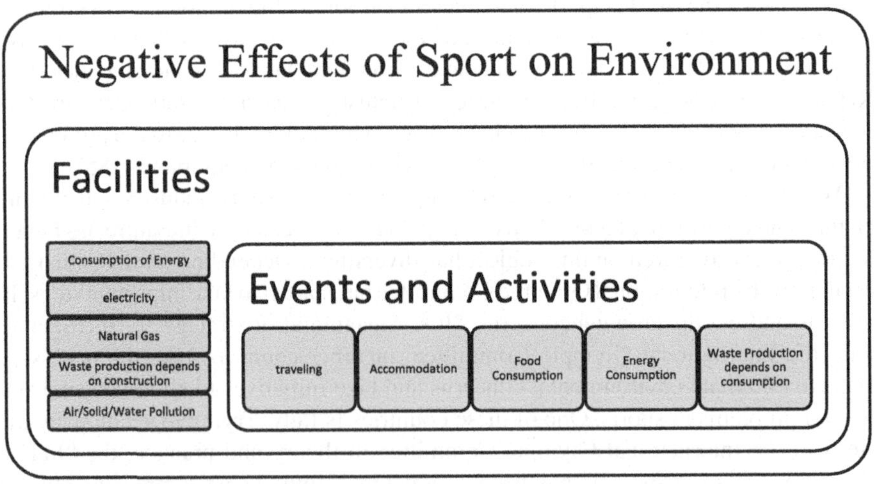

Fig. 2.1 Sources of negative environmental impact in sports. *Source* [75]

Mega sport events such as the FIFA World Cup lead to approximately 3 million kW-h of electricity consumed. What is even more dangerous is that this electricity need is met by fossil fuels [72]. The environmental effects caused by the energy consumption of only one month of organization can be quite destructive. Especially in football, it is worrying that it produces extreme energy consumption, a huge amount of water use and negative environmental outcomes that cause pollution. A study conducted by [74] reveals that the total energy demands of approximately 4,000 stadiums in the European continent, which generally utilize fossil fuel sources, are as high as 40 TWh per year. Khalid and Rogstam [48] state that the average energy consumption of an average sized stadium (55,000 seats) is equal to 10,000 MWh per year (3,600 tCO_2-eq/year). The above-mentioned amounts of energy consumption cause millions of tonnes of carbon dioxide emissions every year, and this amount of emission causes great damage to nature.

Ünal and Bağcı [83] state that the negative environmental affects in sports is growing step by step due to the unplanned consumption of natural resources. Negative environmental impacts are becoming widespread and diversified in the developing sports industry. For example, an international event in 2005 generated approximately 201,000 tonnes of greenhouse gas emissions. In addition, according to calculations, it was reported that participants emitted approximately 60,000 tonnes of carbon and greenhouse gases [1]. According to the results related to waste production in another sports event, it was concluded that 70,000 tonnes of waste were produced [77]. According to the results of scientific research, these effects of sport on the environment pose a threat to sustainable environmental goals.

During sports organizations and events, high amounts of energy are needed for the active and effective operation of the stadiums, along with the increased use of air conditioning, water and electricity due to heating, cooling and lighting of the host facilities [47]. However, as a result of the consumption behaviour of the participants during sporting events, it causes large amounts of waste to be generated, which contributes to the carbon footprint [18].

To give another example from the United States, unlike football, a Super A stadium hosting a bowl event (80,000 seats) consumes up to 10 MWh/h (about 50 MWh over the entire game) during peak energy [10], while Liberia uses less than a third of this amount within its own country [36]. Socially most popular sporting events (Olympic Games, FIFA World Championship, Super Bowl, etc.) are constantly developing the interest of the masses and need larger spaces to host an increasing number of participants (spectators). In order to meet this interest, a significant increase in energy consumption is observed. In sports organizations that appeal to large and large audiences, the energy needs of sports stadiums and facilities are compared with the energy consumption of some small countries in Africa.

However, it can be said that one of the biggest negative environmental effects in sports sector is the choice of transport of active and passive participants [20, 79]. Especially in mega sports events, the travels of people to participate in the organized events are one of the most important emission sources [29, 63]. In addition, besides major sporting events, travelling for local leagues and events is another threat. It is seen that the location of sports events organized in many sports, especially competitive club sports such as swimming, is determined by the relevant institutions, and facilities are located further and further away from population centres, and accordingly, the behaviour of travelling by personal vehicle increases [13]. Hjorthol and Fyhri [42] reveal that most of the organized activities take place outside the immediate environment and that the most typical transportation for these activities is automobile. The human activity that contributes the most to air pollution is individual vehicle and automobile use [7].

According to a study conducted in the United States on the amount of carbon dioxide caused by individual vehicle use, the average carbon dioxide emission per person was calculated as 415,000 kg CO_2, depending on the participants' personal vehicle use to participate in sports activities for an entire season. This amount of emissions is equivalent to one year of electricity used in 62 homes, 88 passenger cars used for one year, or 994,000 miles travelled by an average passenger vehicle [82].

Mass participation in sports, intensive use of raw materials, traffic density especially on competition days, high levels of waste generation and pollution related to all these are some of the important environmental problems [24, 38, 64]. In addition to this, rapid facilitation steps due to increasing demand also associate environmental impacts related to the construction sector with sports [28, 47, 57].

Another cause of negative environmental impact in the sports sector is the waste produced in sports facilities and organizations. For example, in the United States, sports facilities can cause huge levels of greenhouse gas emissions depending on which waste management practices are used [89], and the differentiation of these waste management practices may vary in each region [16]. More sustainable storage or waste collection methods for waste management in sports facilities can greatly reduce greenhouse gas emissions from waste in sports facilities [14]. Also, in addition to direct greenhouse gas emissions from sports facilities, these facilities often remain empty for most of the year, and so the "adaptive reuse" initiative with wider communities in mind contributes to static pollution of unused, unproductive and inaccessible space [34].

Within the sports sector there is a lot of traveling, the construction of new sports facilities, and the production of sports equipment and services [3]. This leads to high amounts of carbon emissions and is considered a significant threat to sustainable environmental goals [58, 80]. For example, the creation of winter sports trails and cable cars that require the destruction of native vegetation (trees, bushes and large rocks) significantly alters the natural composition of the mountains, leading to further damage to the ecosystem [67]. There is an important relationship between the preference of mountainous areas and lands especially for winter sports and the destruction of vegetation. While mountains and mountainous regions have significant water, energy, ecological diversity and natural resources [86], they are also sensitive ecosystems susceptible to soil erosion, landslides and loss of genetic diversity [33, 92]. The serious destruction of these ecosystems due to the construction of facilities for winter sports is one of the negative environmental effects of sports, causing the carbon footprint of sports to grow gradually. Tuppen [81] states that the damage to the environment is ignored due to the economic and commercial activity caused by the ski facilities built.

The infrastructure of countries or cities hosting international sports organizations (airports, stadiums, hotel complexes, etc.) also has a great potential for environmental impact [79]. So much so that the facilities in the host countries and cities most likely affect the ecosystems in their region, It causes the unplanned use of natural resource reserves and can cause high amounts of carbon emissions [15, 46]. There are many factors that reveal the environmental sensitivity of these facilities. Factors such as the energy used in the facilities, the size of the facilities, the geographical location, the structure of the facilities, the presence and type of heating and cooling systems and the production of garbage and wastewater determine the compliance of the facilities with environmental sensitivity [9, 59, 71].

For the continuity of all sports branches, the construction of facilities and the maintenance and repair of existing facilities are of course inevitable. However, wastes generated during construction processes are only one aspect of the problem. The other dimension is the high amount of energy used in these facilities. One of the main sources of carbon emissions is energy consumption. The fact that this energy need is met from fossil fuels is a great risk for the environment and nature.

Many countries want to host major sporting events because it has important effects in the promotion of these events. The duration of the activity of millions of people is going to this country or watching on television [43]. However, in order to host sports facilities in this country, the operation of the facilities must be very low. Many international sports organizations, especially the International Olympic Committee, try to be very sensitive about the negative effects between sports and the environment. In this context, if a country wishes to host a major sporting event [83], it has to fulfil the following, aiming at improving the relationship between sport and the environment:

- Protection of ecosystem and biological structures.
- Planned land natural resource use.
- Preventing soil, air and water surfaces.
- Natural resource and waste management.
- Health and safety measures.
- Reducing environmental disturbances.
- Ensuring the continuity of cultural competence [44].

In the evaluation process of countries' hosting applications, the criteria stated above must be fulfilled. Otherwise, it is a possible possibility that the application will result in a negative result. As stated above, maintaining these rules in sports is very important for the purpose of sustainability and the International Olympic Committee has very serious practices in this regard [56]. As a matter of fact, when hosting is very important for countries, it can be said that countries that are candidates for hosting are now trying to take important steps in these matters. In addition to the International Olympic Committee, the United Nations has started to take certain steps on the negative relationship between sports and the environment.

Alongside the International Olympic Committee, the United Nations Climate Gaming Program draws attention to the conversation between sport and the environment. The United Nations maintained the course of sports in the fight against climate change and included the sports sector in its policies. In these duties, he has assigned different responsibilities to all the responsibilities of sports. The following responsibilities give clues to the chamber of the importance of sports in the fight against climate change:

- Measurement and monitoring of internal greenhouse gas and carbon emissions of the entire sports sector.
- Managing and benefiting from sports in the fight against bilateral change with different cooperation and partnerships.
- To soften the heat distribution and to use sports as a tool by taking advantage of the power of sports distribution [84].

Many spot facilities such as stadiums and basketball halls are very important for the spread of sports and access to sports games. Considering the contribution of sports to health and social development, the presence and number of sports facilities exceed their number. However, the environmental friendliness of these facilities should also be considered while extracting the positive aspects of the necessity of the facilities. The understanding of the facility with the least negative protection should be expanded. And for this, they must take all necessary precautions. This is what should be considered in terms of the sustainability of both sports and the environment [6]. The relationship between sports facilities and the environment is extremely sensitive. It is necessary to give a few examples and understand the seriousness of this issue. With the daily global heating, the last artificial snow is produced on the ski slopes. Therefore, the use of artificial snow to keep clean water consists of destructive effects on the ecosystem with the mixing of wastes into the construction [72]. Apart from skiing, Golf is also an important sports industry. However, the removal of natural vegetation and soil structure during the construction of new golf courses adversely affects the natural balance. Cutting down trees may also cause the extinction of the living things living there [77]. Skiing and golf are just two examples. It is possible to diversify the negative effects of structuring in similar sports because a step taken by children on the natural environment they live in will have consequences. In order to gain a general perspective and get an idea, the relationship between sports and environment put forward by Balcı and Koçak is given in more detail in the Table 2.1 [4].

The table above provides an opportunity for a detailed evaluation. Establishment in sports is not a situation that concerns the sports industry or the construction industry alone. As with all construction steps, there is also an intense relationship with the environment. In the construction processes of the facilities, the natural vegetation is primarily removed. This means the destruction of the habitats of the creatures living there, which also means the destruction of these creatures. However, the opening of fertile agricultural lands or forest areas for construction causes the unplanned and uncontrolled consumption of natural resources. Then, after construction, it causes carbon and greenhouse gas emissions due to high energy and water consumption.

Table 2.1 Negative effects on environment of different sports

Sports	Effects on the environment
Golf	• Opening of agricultural lands • Destroying the vegetation in the area • Continuous water consumption • Inappropriate and uncontrolled pesticide use • Soil and water pollution • Destruction of biodiversity
Ski	• Damage to vegetation • Opening of forest areas to use • Soil deformation due to forest use • Air, soil and water pollution • Intensive chemical use • High amount of waste generation • Threatened wildlife • Decline of wild animal species
Swimming	• A huge amount for water and energy consumption • Unplanned use of chemicals • Failure to control chemical waste
Ice sports	• A huge amount of energy consumption • Liquid waste release • Carbon and greenhouse gas emissions • High consumption of natural resources (water, etc.) • Water and soil pollution • Uncontrolled use of chemicals • Failure to follow up chemical waste
Football	• Air, soil, water and noise pollution • High energy consumption • Participant sourced carbon and greenhouse gas emissions • Carbon emissions from fossil fuels • Intensive waste generation
Car racing	• Use of fertile natural lands for runway construction • Destruction of natural vegetation • Threatened wildlife • Air, water and soil pollution • High energy consumption • High carbon and greenhouse gas emissions
Recreation areas	• Noise pollution in the environment • Unplanned and uncontrolled use of the natural environment • High energy and water consumption • Carbon and greenhouse gas emissions

Chemical use in swimming and ice sports mixes with soil and water, which accelerates pollution. To summarize, the installation and all subsequent processes cause negative environmental effects. Energy use, water consumption, uncontrolled use of chemicals, etc. All processes pose risks to environmental sustainability goals.

The sports sector is one of the biggest industries. New facilities should be constructed and existing facilities should be maintained and repaired. However, environmental policies must be taken into account in all processes. Environmental precautions such as planned and controlled use of natural resources, choosing the right land and establishing the infrastructure for clean energy resources should be taken. In particular, the protection of wildlife and natural vegetation is essential for the continuity of the habitat. Sustainable sports and environmental goals seem possible with environmentally friendly sports policies.

It can be said that the researches revealing the relationship between environment and sports reveal common results. Emphasizing environmental concerns and an environmentally friendly facility approach are emphasized in these studies because the impact power of the sports industry on the masses is quite high. Not only the construction of the facilities, but also the operation of these facilities carries risks for the environment. As stated above, these facilities have high energy consumption (electricity, natural gas, etc.) for many years. It is known that this energy need is met by fossil fuels. Fossil fuels, as the largest source of carbon and greenhouse gas emissions, have devastating effects on the natural environment.

All national and international sports industry stakeholders are trying to draw attention to these problems. For a sustainable environment, it is necessary to reduce all the negative effects of the above-mentioned facilities and activities and to build an environmentally friendly sports mentality. Otherwise, the destructive effects of sports on nature will gradually increase [37].

It is obligatory to control all processes in sports, especially the establishment, and to take steps regarding environmental impacts. Environmental issues need to be addressed in plans and policies. These plans and policies should be implemented both in theory and in practice [25] because all human-based vital activities are closely related to nature and the environment [17]. All behaviours exhibited by people have different levels of influence. It is doubtful that these effects are more negative.

Until recent years, it is not possible to say that the handling and evaluation of negative environmental effects in the sports industry is very efficient when compared to other sectors. Therefore, it can be said that the action plans created against environmental effects are weak and overshadowed by economic concerns and sportive competition. However, with the recent steps taken by international sports organizations (IOC, UEFA, FIFA), it can be stated that environmental problems have begun to find more place in the sports sector. In particular, in the last world cup held in Qatar 2022, sports and environmental sustainability were handled together, and it was seen that environmentally friendly practices were implemented in addition to commercial and sportive plans for this world cup.

Sporting events are indeed a serious challenge for the industry, as large-scale sporting events cause significant adverse environmental impacts, particularly in terms of greenhouse gas emissions, waste generation and energy and water consumption. Due to increasing environmental concerns, some international sports organizations have taken a more proactive stance on environmental sustainability [27]. Pursue not only sports but also energy efficiency, sustainable mobility, reduction of single-use plastics and other important environmental issues. Also, Real Betis Leading European sports clubs such as Balompié and Porto FC are also taking steps towards carbon emissions and the need to maintain carbon neutrality. Another example is Juventus in Italy's well-established teams. From the 2020/2021 season, Juventus plans to plant 200 trees for every goal scored to contribute to the fight against climate change [27].

Sports managers, who care about the relationship between the environment and sports and take concrete steps for this, try to focus on actions that consider environmental sustainability in practices in order to make the eccentric management approach more specific and concrete [70]. From the point of view of the managers involved in sports organizations around the world, the issue of environment emerges as a basic element of general strategic planning such as maintenance and repair of sports facilities, aesthetic concerns and income generation processes [19]. However, at this point, awareness of environmental processes can become an important issue. It is expected that the awareness and knowledge level of sports managers on environmental issues is high in order to deal with environmental processes in sports and recreation organizations in a healthy way [51] because the awareness and knowledge levels of sports managers are quite decisive in the success or failure of the actions [60].

The International Olympic Committee (IOC), which is one of the leading international umbrella sports organizations, emphasizes the importance of environmental issues in the planning and staging of the Summer and Winter Games in line with the environmental sustainability goals in sports and tries to display an environmentally friendly understanding in all organizations [23]. For example, the understanding of sports and environmental sustainability that started with the 1984 Summer Olympic Games held in Los Angeles, the steps taken to mitigate the adverse environmental effects of sports organizations, facilities and all activities related to the organization and sports and environmental sustainability goals came together [41, 50]. One of the concrete steps taken towards sports and environmental sustainability goals is the Tokyo Olympic Games. Of the 37 sports facilities hosting the 2020 Olympic and Paralympic Games in Tokyo, 15 are pre-existing facilities. Environmental concerns are behind promoting the use of existing facilities. By limiting the number of new facilities for the Games, it was aimed to minimize the harmful environmental consequences (and additional cost) associated with construction and land use [55].

Concrete steps taken by the International Olympic Committee have also been met in the member countries and countries have started to take measures to reduce the environmental impacts of sports. Especially, environmental friendly design of football stadiums gained priority [40]. For example, "Thyagaraj Stadium" located in Delhi received a gold rating from the Green Building Council of India in April 2013. Croke Park and Millennium stadiums in Dublin and Cardiff in Europe have been stadiums that have met the "British Sustainability Management Systems Standard" (BS 8901) developed by the British Standards Institute for the competitions and events to be held [11].

The stakeholders of sports are trying to implement different practices due to increasing environmental concerns. In order to get results, a certain time period and experience are needed. However, the results for a sustainable environment and sports are worrisome. There are negative environmental outcomes, especially with the intense competition schedules and increasing installation moves. As pointed out by different studies above, basic negativities such as intensive travels, energy consumption, mass participation and waste production threaten the environment and can have devastating effects on nature.

2.2 Applications to Protect the Nature

Along with increasing environmental concerns, international sports institutions and organizations have also taken important measures. Chief among these institutions is FIFA. There is an intense participation in football, one of the most popular sports branches in the world. Depending on this participation, organization and establishment are quite fast. Especially in countries that will host global tournaments such as the World Cup, new stadiums are being built. In addition, existing stadiums need maintenance and repairs because the criteria set forth by FIFA must be met.

One of the criteria titles determined by FIFA is devoted to the fight against climate change. Completion of the stadiums to be built with an environmentally friendly process is one of the hosting criteria. Especially in the last decade, FIFA has continued to strengthen its environmental protection requirements and programs, covering topics such as climate change, waste management, recycling, sustainable construction and procurement.

2.2.1 The Global Sustainability Assessment System (GSAS)

FIFA is working with the Global Sustainability Assessment System (GSAS) on the environmental friendliness of new stadiums built in host countries and existing stadiums.

The Global Sustainability Assessment System (GSAS) is a system designed to rate and evaluate the environmental impact potential of existing and newly constructed buildings. It is an approach that works with performance-based data, especially applied in the Middle East and North Africa geography. The concept of green building for the environment works in partnership with many respected institutions on a global scale. Georgia Institute of Technology and the Gulf Research and Development Organization (GORD) are among these institutions. The main purpose of GSAS is to protect the natural environment we live in, to use natural resources in a planned and regular manner and to increase individual and social welfare. In this way, it aims to contribute to the realization of sustainable environmental goals [39].

FIFA, one of the umbrella organizations of the sports industry, works in cooperation with GSAS. In this context, FIFA operates GSAS assessment processes in order to reduce the negative environmental impacts related to installations, especially in football. With this cooperation, which includes the evaluation and scoring of the environmental impacts of the stadiums, the rating system is applied during the construction and operation of the stadiums. It requests the host countries that their stadiums pass this rating system.

With its environmentally friendly approach, FIFA takes a determined stance on the construction and operation of stadiums in harmony with nature and the environment. Thanks to this stance, the environmental effects of stadiums can be minimized, and environmental practices can become widespread in the football industry. In order for this process to be carried out correctly, FIFA accepts the certification system determined by GSAS and applies this rating and certification process in all stadiums [30].

The GSAS framework covers eight environmental impact categories under the "Design and Built": civic link, materials, energy, water, indoor environment, site, cultural & economic value and management & operations [39]. As per FIFA requirements, all FIFA World Cup™ stadiums are designed and built following sustainable building standards.

Within the scope of the evaluation system of GSAS, there are three different title certificate programs in which the environmental compatibility of the facilities is discussed and evaluated during the project design process. In this section, all the rating and evaluation steps of the Global Sustainability Assessment System (GSAS) will be explained as they are. And the scoring of the stadiums in the last Qatar 2022 world cup will be looked at. These stages are the Design, Construction and Operations stages and presented (Fig. 2.2).

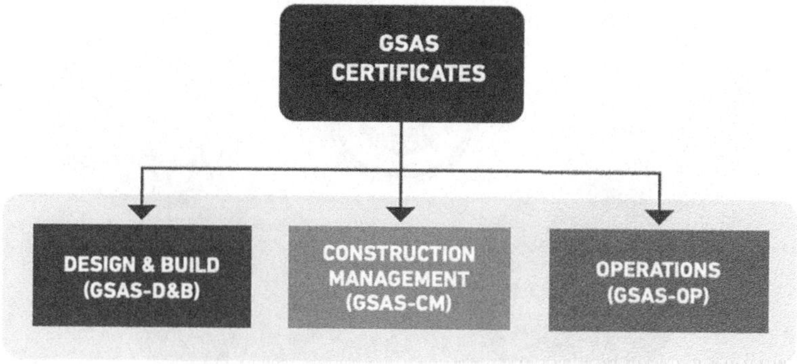

Fig. 2.2 Types of GSAS certification. *Resources* GSAS Technical Guide [39]

2.2.1.1 GSAS Design and Built (D&B) Categories

The GSAS framework covers eight environmental impact categories under the "Design and Built": urban connectivity, energy, indoor environment, water, materials, site, cultural & economic value and management & operations. As per FIFA requirements, all FIFA World Cup™ stadiums are designed and built following sustainable building standards. These categories are presented in the image [39] (Fig. 2.3).

The descriptions of GSAS D&B categories are as follows:

1. Urban Connectivity [UC]: The title of Urban Connection consists of urban elements in cities that are directly or indirectly related to the natural environment, such as construction, transportation infrastructure and cargo areas. It deals with problems such as air, water and soil pollution and traffic density caused by these elements.
2. Site [S]: The site title is mainly concerned with the use of available tools. It is a title that covers the protection of the lands used in this context, the planned construction, site selection and the development of these places.
3. Energy [E]: The title of energy is a title that covers the amount of energy demanded for existing buildings, efficiency in energy use and distribution and carbon and greenhouse gas emissions due to fossil energy sources.

Fig. 2.3 GSAS D&B categories

4. Water [W]: The title water is a title that covers the amount of water consumption in cities and buildings and the compatibility of this water supply and distribution processes with the environment and focuses on environmental impact factors.
5. Materials [M]: Materials category; processing them, addressing the possibilities of reuse, constructing and ultimately destroying the production and distribution processes.
6. Indoor Environment [IE]: The indoor environment title includes the indoor air, sound and lighting quality in buildings. In addition, it includes the evaluation of the environmental relationship with the energy used in indoor spaces.
7. Cultural & Economic Value [CE]: The process refers to the support and cultural protection of the national economy.
8. Management & Operations [MO]: It covers the design, operation, management and all operational processes of these buildings [39].

The evaluation and scoring process includes scoring from −1 to 3. −1 is considered a negative rating, while 3 represents the ideal level. On the GSAS scoring scale, −1 means unacceptable level and the environmental impact value is quite high.

The project and construction scoring and evaluation system includes six ratings and certification levels to determine the environmental impact of buildings or facilities. Each certification level consists of one star, while the highest certification grade consists of six stars (refer Fig. 2.4) [39].

SCORE	RATING
$X < 0$	Certification Denied
$0.00 \leq X \leq 0.50$	★
$0.50 < X \leq 1.00$	★★
$1.00 < X \leq 1.50$	★★★
$1.50 < X \leq 2.00$	★★★★
$2.00 < X \leq 2.50$	★★★★★
$2.50 < X \leq 3.00$	★★★★★★

Fig. 2.4 The project and construction scoring and evaluation system

2.2.1.2 GSAS-Constriction Management (CM) Categories

The criteria of GSAS-Constriction Management consists of eight headings. Each chapter includes the evaluation, follow-up and measurement of the practices in the construction processes. In this way, it is ensured that the people responsible for the construction of the construction act with an environmentally friendly approach. These eight main headings are subdivided in the next step and focus on identifying and measuring problems caused by individuals (refer Fig. 2.5).

1. Urban Considerations [UC]: Urban title focus on water pollution, drainage and storm water works and traffic problems, which are infrastructure components.
2. Site [S]: The site title focuses on natural resource conservation topics such as soil, supply and biodiversity in areas where construction takes place.
3. Energy [E]: The title of energy focuses on the harmony of the temporary buildings created during the construction process and the energy used in the construction activities with the environment. In this process, issues such as the amount of energy consumption and its sources are discussed.

Fig. 2.5 GSAS-CM categories

4. Water [W]: The water title focuses on processes related to the water consumed both inside and around the building. In these processes, clean water consumption and consumption amounts are evaluated in terms of compatibility with the environment.
5. Materials [M]: The title of the material focuses on the effects of construction waste on the environment while monitoring the collection, storage and disposal of wastes generated during the construction process. Especially since there is a lot of waste production in the construction process, the relationship of the process with the environment is controlled.
6. Outdoor Environment [OE]: The outdoor title focuses on sound, dust, noise, vibration, odour, VOCs emissions and visual pollution caused by the construction process. As a part of the construction process, air, soil and water pollution are evaluated and monitored in terms of their environmental effects.
7. Socio-Cultural Dimension [SD]: The heading of socio-cultural dimensions focuses on the preservation of historical remains, archaeological finds, sites and heritage sites. It is accepted that the construction sites may be in historical, cultural and archaeological areas, necessary controls and inspections are carried out and it is aimed to protect the remains accepted as cultural heritage.
8. Management & Operations [MO]: The heading of Management and operations includes a comprehensive assessment and covers: Targeted criteria are evaluated in three different construction phases; Activation/Foundation Phase, Infrastructure and Superstructure Phase, and Finishing Phase.

The intended considerations address and evaluate the three different processes of the construction phase. The activation/establishment step is the construction of the superstructure and infrastructure, and finally the finalization/finishing step. All these steps are evaluated separately in the whole process and as a result of the evaluation, scoring is made for each step. The building is given a certificate according to scoring. Scoring process and steps are done as follows (refer to Fig. 2.1). The minimum value for any construction project is 0.5 points. The equivalent of this fee is that the building is within the scope of class D. The maximum value for the building is 2.5 points, which is in the scope of class A (refer to Table 2.2) [39].

Table 2.2 GSAS-CM tabulated certification scores and ratings

Score	Rating
$X \leq 0.5$	Certification denied
$0.5 \leq X < 1.0$	Class D
$1.0 \leq X < 1.5$	Class C
$1.5 \leq X < 2.0$	Class B
$2.0 \leq X < 2.5$	Class A
$X \geq 2.5$	Class A*

2.2.1.3 GSAS-Operations (OP) Categories

The GAS-OP evaluation system consists of six titles. Each chapter focuses on the environmental impacts of the operational processes of buildings. It also includes suggestions on how to reduce existing negative impacts. These titles make definitions by evaluating individual-induced negativities (refer Fig. 2.6).

Detailed explanations of the GSAS-OP titles are given below:

1. Energy [E]: The energy title focuses on consumption and the relationship between energy supply and the environment. The current negative environmental effects of energy consumption and supply are monitored and evaluated and necessary measurements are made.
2. Water [W]: The water title focuses on consumption and supply processes, as in the energy section, and includes evaluations and measurements regarding its relationship with the environment. The relationship between natural resource consumption and the environment is discussed.

Fig. 2.6 GSAS-OP categories

3. Indoor Environment [IE]: The indoor environment title includes the indoor air, sound and lighting quality in buildings. In addition, it includes the evaluation of the environmental relationship with the energy used in indoor spaces.
4. Waste Management [WM]: The title of waste management is a system that evaluates sub-titles such as reducing the amount of potential solid waste, recycling and reuse. It especially focuses on the collection, recycling and disposal of solid wastes with an environmentally friendly approach.
5. Facility Management [FM]: The title of facility management focuses on the management of all kinds of buildings built with an understanding of sustainability. This understanding aims to develop plans, programs and strategies. The primary goal is to operate buildings that will survive for many years after the construction phase in an environmentally friendly manner.
6. Environmental Policy & Awareness [EPA]: Environmental policy and awareness title focuses on the sustainable relationship between building and environment. To this end, it aims to organize initiatives and events to increase and encourage awareness [39].

The title of GSAS operations, which includes a quantitative and qualitative assessment, means that the requirements for 0 points are not met. However, a score of 3 out of 1 means improving and environmentally friendly buildings. Five certification scales are used in the evaluation and scoring of buildings and facilities under the Operations heading. These certification scales are Bronze, Silver, Gold, Platinum and Diamond. A Bronze certificate is the lowest level, while Diamond is the highest (refer Table 2.3) [39].

GSAS is a system that scores stadium construction and operation processes in sports and especially football. In the short term, it seems that it will mediate in reducing the negative environmental effects caused by the establishment in sports. Countries wishing to host international tournaments must build facilities that meet the GSAS criteria. With the evaluation and scoring, the environmental effects of international organizations such as the world cup can also be followed, and comparisons with other organizations can be made. The most recent example of this situation is the FIFA 2022 Qatar World Cup organization.

The cooperation between FIFA and the Global Sustainability Assessment System (GSAS) became concrete with the last Qatar 2022 world cup organization. There are 8 stadiums in the host country for the competitions to be played. The environmental

Table 2.3 GSAS-OP tabulated certification scores and ratings

Score	Rating
$X \leq 0.5$	Certification denied
$0.5 \leq X < 1.0$	Bronze
$1.0 \leq X < 1.5$	Silver
$1.5 \leq X < 2.0$	Gold
$2.0 \leq X < 2.5$	Platinum
$X \geq 2.5$	Diamond

impact assessments of these stadiums were made through the Global Sustainability Assessment System standards. As a result of the evaluations, the sustainability certificates and degrees prepared for 8 stadiums are presented visually in the Table 2.4 [30].

Hosting the World Cup 2022, Qatar has built sustainable stadiums and the necessary areas for the organization in terms of environmental sustainability criteria. It can be said that these facilities, which include the combination of economic and environmental sustainability, are very important for the future of sports and societies. It can be accepted that the technology and infrastructure systems used are also a reference for other countries and other organizations. The following minimum certification standards are being applied for all FIFA World Cup 2022™ stadiums:

- GSAS Design & Build: Certification Requirement: 4 Stars (maximum = 5 Stars).
- GSAS Construction Management: Certification requirement: Class A (maximum = Class A*).
- GSAS Operations: Certification requirement: Gold (Platinum and Diamond supersede Gold) [39].

It applies the GSAS evaluation system in all international organizations organized by FIFA. In this way, it aims to contribute to the construction and operation of stadiums with an environmentally friendly approach. However, it also attaches particular importance to the sustainability of stadiums, for example, after the world cup tournaments. To give an example specific to Qatar, the 8 stadiums that host it are not up-to-date but have old building system plans. Under this plan, it is possible to reuse the seats in the stadiums after the World Cup. Among these stadiums, the seating capacity of Al Janoub, Ahmad Bin Ali, Al Bayt Stadium, Al Thumama and Education City stadiums can be reduced by half after the tournament and the removed seats can be used in other facilities. Therefore, environmental impacts will be reduced thanks to materials with a long service life. Sustainable football and environmental goals can be achieved thanks to these stadiums, which are the most up-to-date examples of sustainable and environmentally friendly stadium understanding [31].

Table 2.4 Certification of stadiums in qatar

Stadium	Design & build	Construction management	Operations
Al Bayt	5 Stars	A*	Platinum
Al Janoup	4 Stars	A*	Gold
Al Thumama	5 Stars	A*	Gold
Education City	5 Stars	A*	Gold
Khalifa International	4 Stars	A*	Gold
Ahmed Bin Ali	4 Stars	A*	Gold
Lusail	5 Stars	A*	Not targeted
Stadium 974	5 Stars	A*	Not targeted

Together with FIFA, it implements written documents and strategy policies on the relationship between environment and sports in the International Olympic Committee and the United Nations. In particular, the sustainability strategy document of the International Olympic Committee is very important in terms of ensuring the balance between sports and the environment. The relevant strategy document is presented in the appendices of this study.

2.2.2 *As a Solution Green Stadiums*

Constructing sports facilities that are sensitive to nature and the environment, low energy consumption within these facilities, facilitating access to existing sports facilities, disseminating environmentally friendly practices such as bicycle use and ensuring the continuity of sports activities as well as a sustainable environment can contribute to individual and social health and the living environment [62]. Green stadium ideas are at the forefront of the steps taken in response to sports and environmental sustainability concerns. In this part of the research, the evaluation of environmentally friendly stadiums designed, built and operated in recent years and some examples of environmentally friendly stadiums are included.

Environmentally sensitive and green concept stadiums represent environmental awareness together with new generation technologies. Especially stadium models with reduced negative environmental impacts are very important initiatives for the sustainability of both sports and the environment [35].

2.2.2.1 Forest Green Rovers

The team generates a significant portion of the energy used in their home stadium, The New Lawn in Nailsworth, by installing solar cells and is the first football club to adopt an all-vegan diet, an effective way to combat climate change [35].

2.2.2.2 Kaohsiung World Stadium—Taiwan

It is known that the total capacity of this stadium is 55,000. It is used by the Taipei national team, especially for international competitions. The most distinctive feature that makes this stadium stand out in the world is that it fully complies with the sustainability criteria, according to the reports obtained. Another outstanding feature of this stadium is the use of materials suitable for sustainability during the construction process. And it is covered with panels that can reach 9,000 photovoltaic standards with high electrical energy potential. Since it mostly hosts national competitions, the intensity of use of this stadium is not very high. Therefore, a high amount of energy is accumulated thanks to the panels. This accumulated energy is sold to electricity distribution companies. In this way, the stadium has the potential to generate

Fig. 2.7 Kaohsiung world stadium

economic value even if it is not used. On the days and times of the competitions, 4/3 of the stadium's energy needs are met with the energy obtained from these panels. In addition, the landscaping around the stadium is compatible with the environment and free from negative effects [8] (Fig. 2.7).

2.2.2.3 Mercedes Benz Arena—Atlanta

Today, stadiums stand out with their technological infrastructure and environmentally friendly features. One of these stadiums is the Mercedes Benz Arena in the United States. Due to its features, it is one of the most successful stadiums in terms of sustainability. The stadium includes a nature-friendly and environmentally friendly approach in all management and operation processes, starting from the construction phase. The stadium is extremely efficient in energy and natural resource consumption. It has the potential to provide energy savings of close to 30% in general usage processes. In addition, the stadium management offers fans environmentally friendly transportation preferences. So much so that it carries out activities for the use of bicycles for the participants until 2 h before the competitions [8] (Fig. 2.8).

Fig. 2.8 Mercedes Benz Arena

2.2.2.4 Amsterdam Arena

Amsterdam Arena is one of the rare structures in the world where both solar energy and wind energy are used simultaneously for electricity needs. The stadium infrastructure was built almost entirely with an environmentally friendly approach. This stadium, which stands out with its applications beyond the age, such as the use of electric vehicles and escalators used in energy production, also makes a difference with its energy storage systems. Electric vehicle batteries are used for energy storage and provide a significant income from this. There is a rainwater harvesting system in the stadium, which also has practices related to the protection of basic natural resources such as water. Rainwater transferred from the stadium roof is stored and this system mediates the protection of fresh water resources. The stadium management, which also includes environmental practices for the fans, encourages the use of electric vehicles. In addition, the fans who come to the competition benefit from public transportation vehicles with discounts through their competition tickets. In this way, individual vehicle usage and emission rates are tried to be reduced [22] (Fig. 2.9).

2.2.2.5 New Tottenham Hotspur Stadium—London

Although the zero emission policy related to the sports sector is very ambitious, Tottenham Hotspur Stadium is another stadium that stands out with its zero emission

Fig. 2.9 Amsterdam ArenA

concept in the construction and management process. The use of completely renewable energy in the process of meeting all the energy needs of this stadium gives this stadium the title of the most environmentally friendly building. While the energy need of the stadium is completely met through solar energy panels, harvest water is used in areas such as toilets. According to all these features, it can be said that the negative environmental effects of the stadium are almost zero [8] (Fig. 2.10).

Fig. 2.10 New Tottenham Hotspur Stadium

2.2.2.6 Allianz Riviera—Nice

Allianz Riviera is the first stadium built with an environmentally friendly approach on the European continent. This stadium, which is very restrictive in terms of energy saving and use, is the most important example of the relationship between nature and the sports sector. While the energy obtained from wind energy is used for the stadium need, it also meets the energy needs of the region. In addition, the protection of clean water resources with the rainwater harvesting system is another feature of this stadium. Perhaps the most important feature of the stadium is that it has a geothermal system. In this way, the building can be easily heated with zero environmental impact, and no extra energy is required for this. In this facility, where the harmony between nature and the stadium is emphasized, 100% environmental practices such as energy saving, energy supply to the environment and protection of clean water resources come to the fore. In this way, it contributes to financial sustainability as well as environmental sustainability [8] (Fig. 2.11).

The concept of environmentally friendly stadiums is becoming more and more widespread. Especially with the use of clean energy sources in stadiums, while the effects on the environment are reduced, it also contributes to the financial sustainability of the clubs. With the decrease in energy consumption, there is a decrease in financial expenses. However, the understanding of green and environmentally friendly stadiums and sports facilities should be widespread and these practices should be started in other sports.

Fig. 2.11 Allianz Riviera

References

1. F. Ahmed, L. Pretorius, Mega-events and environmental impacts: the 2010 FIFA World Cup in South Africa. Alternation **17**(2), 274–296 (2010)
2. A. Athanassiadis, R.H. Crawford, P. Bouillard, Overcoming the 'black box' approach of urban metabolism, in eds. by R.H. Crawford, A. Stephan *Living and Learning: Research for a Better Built Environment* (4th edn. The University of Melbourne, pp. 574–556
3. K. Babiak, S. Trendafilova, CSR and environmental responsibility: motives and pressures to adopt green management practices. Corp. Soc. Responsib. Environ. Mgmt. **18**, 11–24 (2011). https://doi.org/10.1002/csr.229
4. V. Balcı, F. Koçak, Spor ve rekreasyon alanlarının tasarımında ve kullanımında çevresel sürdürülebilirlik. Spor ve Performans Araştırmaları Dergisi **5**(2), 46–58 (2014)
5. D. Bălteanu, D. Dogaru, Gepraphical perspectives on human-environment relationships and anthropic pressure indicators. Rom. J. Geogr. **55**(2), 69–80 (2011)
6. M. Barghchi, D. Omar, M.S. Aman, Sports facilities in urban areas: trends and development considerations. Pertanika J. Soc. Sci. Human. **18**(2), 427–435 (2010)
7. J. Barkenbus, Eco-driving: an overlooked climate change initiative. Energy Policy **38**, 762–769 (2009)
8. R. Belardinelli, The 5 best green stadium in the world (2021). https://www.nssmag.com/en/sports/25624/5-migliori-stadi-sostenibili. Accessed 27 Mar. 2023
9. P. Bohdanowicz, I. Martinac, Determinants and benchmarking of resource consumption in hotels—Case study of Hilton International and Scandic in Europe. Energy Build. **39**(1), 82–95 (2007)
10. J. Breech, Cowboys stadium uses more electricity than liberia. CBS Sport (2013). www.cbssports.com. Accessed 14 Mar. 2023
11. British Standards Institution, Standards (2013). http://www.bsigroup.com. Accessed 24 Mar. 2023
12. H. Bulkeley, Cities and the governing of climate change. Ann. Rev. Environ. Resour. **35**(1), 229–253 (2010)
13. K. Bunds, J. Casper, Sport, physical culture, and the environment: an introduction. Sociol. Sport J. **35**(1), 1–7 (2018). https://doi.org/10.1123/ssj.2018-0007
14. M.T.I. Cabaraban, M.V. Khire, E.C. Alocilja, Aerobic in-vessel composting versus bioreactor landfilling using life cycle inventory models. Clean Technol. Environ. Policy **10**(1), 39–52 (2008)
15. H. Cantelon, M. Letters, The making of the IOC environmental policy as the third pillar of the Olympic movement. Int. Rev. Sociol. Sport **35**(3), 294–308 (2000)
16. A.J. Caplan, T.C. Grijalva, P.M. Jakus, Waste not or want not? A contingent ranking analysis of curb side waste disposal options. Ecol. Econ. **43**(2–3), 185–197 (2002)
17. J.M. Casper, M.E. Pfahl, Environmental behaviour frameworks of sport and recreation undergraduate students. Sport Manag Educ J **6**, 8–20 (2012)
18. J.M. Casper, M.E. Pfahl, Environmental sustainability practices in US NCAA Division III athletics departments (2015). Int. J. Event Manag. Res. 10
19. J.M. Casper, M.E. Pfahl, M. McSherry, Athletics department awareness and action regarding the environment: a study of NCAA athletics department sustainability practices. J. Sport Manag. **26**(1), 11–29 (2012)
20. C. Chard, C. Mallen, Examining the linkages between automobile use and carbon impacts of community-based ice hockey. Sport Manag. Rev. **15**(4), 476–484 (2012). https://doi.org/10.1016/j.smr.2012.02.002
21. R. Čiegis, M. Vijūnas, *Sustainable Banking: theoretical and Practical Aspects. Sustainable Development: theory and Practice* (Vilnius University Press, 2014)
22. Climate Action, The 5 most sustainable sports venues in the World (2018). https://www.climateaction.org/news/the-5-most-sustainable-sports-venues-in-the-world. Accessed 27 Mar. 2023
23. A. Collins, A. Flynn, M. Munday, A. Roberts, Assessing the environmental consequences of major sporting events: the 2003–04 FA Cup Final. Urban Stud. **44**, 457–476 (2007)

24. A. Collins, C. Jones, M. Munday, Assessing the environmental impacts of mega sporting events: two options? Tour. Manag. **30**(6), 828–837 (2009). https://doi.org/10.1016/j.tourman.2008.12.006
25. S. Cornelissen, M. Wolfgang, On the political economy of feel good' effects at sport mega-events: experiences from FIFA Germany 2006 and prospects for South Africa 2010. Alternation **17**(2), 96–120 (2010)
26. L. Cucek, J.J. Klemes, Z. Kravanja, A review of footprint analysis tools for monitoring impacts on sustainability. J. Clean. Prod. **34**(2012), 9–20 (2012)
27. T. Daddi, F. Iraldo, R. Puccetti, N. Todaro, The environmental management of professional football events: a survey of football managers. Life Tackle Project Report. Co-funded by the European Commission under the LIFE Programme (2021)
28. M. Dolf, P. Teehan, Reducing the carbon footprint of spectator and team travel at the University of British Columbia's varsity sports events. Sport Manag. Rev. **18**(2), 244–255 (2015). https://doi.org/10.1016/j.smr.2014.06.003
29. E. Du Preez, E. Heath, Determining the influence of the social versus physical context on environmentally responsible behaviour among cycling spectators. J. Sport Tour **20**, 123–143 (2016)
30. FIFA, Sustainable stadiums and offices (2002a). https://publications.fifa.com/en/sustainability-report/environmental-pillar/sustainable-building/sustainable-stadiums-and-offices/. Accessed 27 Mar. 2023
31. FIFA, Legacy plans for tournament sites (2022b). https://publications.fifa.com/en/sustainability-report/economic-pillar/post-tournament-sites-and-asset-use/legacy-plans-for-tournament-sites/. Accessed 27 Mar. 2023
32. FIFA—Fédération Internationale de Football Association, Sustainability Report of the 2014 FIFA World Cup Brazil™ (2014)
33. M. Fischer, S. Wipf, Effect of low-intensity grazing on the species-rich vegetation of traditionally mown subalpine meadows. Biol. Cons. **104**(1), 1–11 (2002)
34. Flaccavento, A. (2016). *Building A Healthy Economy from the Bottom Up: harnessing Real-World Experience for Transformative Change* (University Press of Kentucky)
35. Football Ground Guide, Eco-friendly football stadiums: which are the greenest? (2023). https://footballgroundguide.com/news/eco-friendly-football-stadiums.html. Accessed 27 Mar. 2023
36. P. Foster, How football is driving energy efficiency and sustainability targets, love energy savings (2014). www.loveenergysavings.com. Accessed 14 Mar. 2023
37. D. Getz, Event tourism: definition, evolution, and research. Tour. Manag. **29**(3), 403–428 (2008). https://doi.org/10.1016/j.tourman.2007.07.017
38. T.J. Grant, Green monsters: examining the environmental impact of sport stadiums. Villanova Environ. Law J. **25**(1), 148–176 (2014)
39. GSAS Trust, GSAS Technical Guide 2019. GSAS Publication Series (2019). https://gsas.gord.qa/wp-content/uploads/2021/10/GSAS-Technical-Guide_2019.pdf. Accessed 27 Mar. 2023
40. A. Hershkowitz, Play green: how big league sports is leading the environmental charge. Huffington Post. (2013). http://www.huffingtonpost.com. Accessed 24 Mar. 2023
41. H. Hiller, Assessing the impact of mega-events: a linkage model. Curr. Issue Tour. **1**(1), 47–57 (1998)
42. R. Hjorthol, A. Fyhri, Do organized leisure activities for children encourage car-use? Trans. Res. Part A: Policy Pract. **43**(2), 209–218 (2009)
43. K.H. Hu, F.H. Chen, G.H. Tzeng, Evaluating the improvement of sustainability of sports industry policy based on MADM. Sustainability **8**(7), 606 (2016)
44. International Olympic Committee, Manuel on sport and the environment (2009). https://www.olympic.org/en/Home/Documents/Documents/Olympism%20in%20Action/Olympism%20in%20Action/Environment/Manual%20on%20Sport%20and%20the%20Environment. Accessed 24 Mar. 2023
45. International Olympic Committee, IOC sustainability strategy (2017). http://extrassets.olympic.org/sustainability-strategy/executivesummary/1-1. Accessed 14 Mar. 2023

46. C. Jones, Assessing the environmental impact of a major sporting event. Tour. Econ. **14**, 343–360 (2008)
47. T.B. Kellison, M.J. Mondello, Organizational perception management in sport: The use of corporate pro-environmental behaviour for desired facility referenda outcomes. Sport Manag. Rev. **15**, 500–512 (2012)
48. W. Khalid, J. Rogstam, Energy usage prediction model comparing outdoor vs. indoor ice rinks. Energy Build. **67**, 195–200 (2013)
49. N.H. Larsen, E.G. Hertwich, The case for consumption-based accounting of greenhouse gas emissions to promote local climate action. Environ. Sci. Policy **12**(7), 791–798 (2009)
50. H. Lenskyj, *Best Olympics Ever? The Social Impacts of Sydney 2000* (SUNY Press, 2002)
51. M. Lombardi, P. Pazienza, R. Rana, The EU environmental-energy policy for urban areas: the Covenant of Mayors, the ELENA program and the role of ESCos Energ. Policy **93**, 33–40 (2016)
52. P. Macnaghten, Embodying the environment in every-day life practices. Sociol. Rev. **51**, 63–84 (2003). https://doi.org/10.1111/1467-954X.00408
53. S. Manfredi, D. Tonini, T.H. Christensen, H. Scharff, Landfilling of waste: accounting of greenhouse gases and global warming contributions. Waste Manag. Res. **27**, 825–836 (2009)
54. M.E. Mann, L.R. Kump, *Dire Predictions: understanding Global Warming* (DK Publishing, 2008)
55. A. Martin, Tokyo promises to meld technology, tradition with 2020 Olympics. Wall Street J. (2013). http://online.wsj.com. Retrieved 24 Mar. 2023
56. A. McCool, Sustainable tourism: guiding fiction, social trap or path to resilience? in ed. by T.V. Singh Challenges in tourism research channel view, 1st edn. (Bristol, 2015), pp. 224–234
57. B. McCullough, T. Kellison, Go green for the home team: sense of place and environmental sustainability in sport. J. Sustain. Educ. **11**, 1–14 (2016)
58. B.P. McCullough, M. Pfahl, S.N. Nguyen, The green waves of environmental sustainability in sport. Sport Soc. **19**, 1040–1065 (2016). https://doi.org/10.1080/17430437.2015.1096251
59. A.V. Michailidou, C. Vlachokostas, N. Moussiopoulos, D. Maleka, Life cycle thinking to assess the environmental impacts of tourism activity for a greek tourism destination. J. Clean. Prod. **111**, 499–510 (2015)
60. A. Nicolaides, The implementation of environmental management towards sustainable universities and education for sustainable development as an ethical imperative. Int. J. Sustain. High. Educ. **7**, 414–424 (2006). https://doi.org/10.1108/14676370610702217
61. C. Özsoy, Düşük karbon ekonomisi ve Türkiye'nin karbon ayak izi. Hak İş Uluslararası Emek ve Toplum Dergisi **4**(9), 198–215 (2015)
62. C.A. Öztekin, Sürdürülebilir yaşam rehberi. J. Manag. Theory Pract. Res. **2**(2), 178–181 (2020)
63. R.P.T. Pereira, M.V.O. Camara, G.M. Ribeiro, V. Filimonau, Applying the facility location problem model for selection of more climate benign mega sporting event hosts: a case of the FIFA World Cups. J. Clean. Prod. **159**, 147–157 (2017)
64. M. Pfahl, The Environmental awakening in sport. Solut. J. **4**, 67–76 (2013)
65. A. Pittock, *Climate Change: turning Up the Heat* (CSIRO Publishing, 2005)
66. A. Ross, The UK approach to delivering sustainable development in government: a case study in joined-up working. J. Environ. Law **17**(1), 27–49 (2005)
67. E. Ruth-Balaganskaya, K. Myllynen-Malinen, Soil nutrient status and revegetation practices of downhill skiing areas in finnish lapland-a case study of Mt. Ylläs. Landsc. Urban Plan. **50**(4), 259–268 (2000)
68. J. Ruževičius, M. Dapkus, Methodologies for calculating the carbon footprint of small organizations. Calitatea **19**(167), 112–117 (2018)
69. R. Sarkar, Public policy and corporate environmental behaviour: a broader view. Corp. Soc. Responsib. Environ. Manag. **15**, 281–297 (2008)
70. M.L. Sartore-Baldwin, B. McCullough, Equity-based sustainability and ecocentric management: creating more ecologically just sport organization practices. Sport Manag. Rev. **21**(4), 391–402 (2018). https://doi.org/10.1016/j.smr.2017.08.009

71. C. Scheuer, G.A. Keoleian, P. Reppe, Life cycle energy and environmental performance of a new university building: modelling challenges and design implications. Energy Build. **35**(10), 1049–1064 (2003)
72. C. Schmidt, Putting the earth in play: environmental awareness and sports. Environ Health Perspect **114**(5), A286–A295 (2006). https://doi.org/10.1289/ehp.114-a286
73. P. Shrivastava, The role of corporations in achieving ecological sustainability. Acad. Manag. Rev. **20**(4), 936–960 (1995)
74. T. Smulders, *Green stadiums: as Green As Grass (Master's thesis)* (University of Utrecht, 2012)
75. F. Sorrentini, The environmental impact of sports activities. Good practices for sustainability: the case of golf. Documenti Geografici **2**, 219–237 (2021). https://doi.org/10.19246/DOCUGEO2281-7549/202102_15
76. W. Steffen, J. Rockström, K. Richardson, T.M. Lenton, C. Folke, D. Liverman, C.P. Summerhayes, A.D. Barnosky, S.E. Cornell, M. Crucifix, J.F. Donges, I. Fetzer, S.J. Lade, M. Scheffer, R. Winkelmann, H.J. Schellnhuber, Trajectories of the earth system in the anthropocene. Proc. Natl. Acad. Sci. U.S.A. **115**(33), 8252–8259 (2018). https://doi.org/10.1073/pnas.1810141115
77. L. Thibault, Globalization of sport: an inconvenient truth. J. Sport Manag. Hum. Kinet. **23**, 1–20 (2009)
78. S. Trendafilova, B. McCullough, M. Pfahl, S.N. Nguyen, J. Casper, M. Picariello, Environmental sustainability in sport: current state and future trends. Glob. J. Adv. Pure Appl. Sci. **3**(1), 9–14 (2014)
79. S. Triantafyllidis, R.J. Ries, K. Kaplanidou, Carbon dioxide emissions of spectators' transportation in collegiate sporting events: comparing on-campus and off-campus stadium locations. Sustainability **10**(1), 241 (2018). https://doi.org/10.3390/su10010241
80. S.A. Triantafyllidis, Carbon dioxide emissions research and sustainable transportation in the sports industry. J. Carbon Res. **4**(4), 57 (2018). https://doi.org/10.3390/c4040057
81. J. Tuppen, The restructuring of winter sports resorts in the French Alps: problems, processes and policies. Int. J. Tour. Res. **2**(5), 327–344 (2000)
82. U.S. Environmental Protection Agency (2016). http://tinyurl.com/j3glbf8. Accessed 22 Mar. 2023
83. H. Ünal, E. Bağcı, Sports organizations in the light of environmental sustainability and ecologic footprint. J. Hum. Sci. **14**(3), 3006–3021 (2017)
84. United Nations Climate Changes. https://unfccc.int/climate-action/sectoral-engagement/sports-for-climate-action. Accessed 24 Mar. 2023
85. United Nations Environment Program, Sport and the environment (2022). https://www.unep.org/news-and-stories/press-release/sports-organisations-commit-safeguarding-nature-under-new-framework. Accessed 14 Mar. 2023
86. United Nations, *Agenda 21: the United Nations Programme of Action from Rio* (United Nations Publications, 1993)
87. United Nations, *World Urbanization Prospects: the 2014 Revision, Highlights* (ST/ESA/SER.A/352) (New York, USA, 2014)
88. United States Environmental Protection Agency, Greenhouse Gas Emissions and removals (2022). https://www.epa.gov/ghgemissions. Accessed 12 Apr. 2023
89. K.A. Weitz, S.A. Thorneloe, S.R. Nishtala, S. Yarkosky, M. Zannes, The impact of municipal solid waste management on greenhouse gas emissions in the United States. J. Air Waste Manag. Assoc. **52**(9), 1000–1011 (2002)
90. R. Welford, C. Chan, M. Man, Priorities for corporate social responsibility: a survey of businesses and their stakeholders. Corp. Soc. Responsib. Environ. Manag. **15**, 52–62 (2007)
91. D. Williamson, G. Lynch-Wood, J. Ramsay, Drivers of environmental behaviour in manufacturing SMEs and the implications for CSR. J. Bus. Ethics **67**, 317–330 (2006)
92. P. Wipf, J. Xiao, J. Jiang, N.A. Belikova, V.A. Tyurin, M.P. Fink, V.E. Kagan, Mitochondrial targeting of selective electron scavengers: synthesis and biological analysis of hemigramicidin—TEMPO conjugates. J. Am. Chem. Soc. **127**(36), 12460–12461 (2005)

Chapter 3
Legal Relationship Between Sport and Environment in Lithuania

Climate change and global warming emerge as an internationally recognized problem as current problems that cause adverse effects on humanity and mainly originate from greenhouse gas (GHG) emissions from both industrial and human-based activities [25]. Studies on the negative environmental impacts of societies show that current unsustainable consumption and production habits are responsible for a wide variety of environmental problems such as climate change, loss of biodiversity and depletion of resources, and in many cases current global consumption levels are at levels that threaten the biological capacity of the planet [3]. In this context, the need to mitigate greenhouse gas (GHG) emissions to mitigate climate change and its effects has become a primary issue in terms of economic, environmental and energy policies [6].

The global climate change problem affects all societies and geographies. The protection of the natural environment and a sustainable environment are among the priority policy areas of all institutions and organizations on a global scale. Increasing social life activities contribute to keeping this issue up to date. Social mobility areas such as intensive consumption, increased use of personal vehicles and the need for continuous energy trigger carbon and greenhouse gas emissions and increase them day by day. In addition, with the concentration of human life especially in cities, construction is accepted as the source of basic problems such as noise and environmental pollution. Therefore, in order to reduce the effects of all these processes, which are an environmental threat, all institutions and organizations create environmental policies, and governments create laws, plans and policies.

The relationship between environment and sustainability, which has become a scientific and theoretical field of study in practice and theory, has been associated with the sports sector and the opinion that important steps should be taken in this regard comes to the fore. Especially, the second part of this study has been designed to reveal this claim, and references are made to environmental problems in sports.

Lithuania is one of the countries where the relationship between sports and the environment is handled with seriousness and sensitivity. Almost 65% of the country, which stands out with its nature, is covered with forest areas. In addition, it is possible

A. Atalay and B. Švagždienė, *Achieving Environmental Sustainability in the Sports Sector*, SpringerBriefs in Applied Sciences and Technology,
https://doi.org/10.1007/978-3-031-37401-2_3

to see many parks and forest areas in different urban centres of the country. However, the increasing production and consumption curve has led Lithuania to be more sensitive about the protection of the natural environment. As in the whole world, the mobility experienced in the sports sector in Lithuania has made it necessary to consider the relationship between sports and the environment within the legal framework. It is possible to see global problems such as continuity in the transportation sector, high energy consumption and intensive waste production due to the intense sports calendar in Lithuania. Therefore, Lithuanian Governments have removed all legal uncertainties in order to protect the natural environment and to establish the understanding of environmental sustainability in sports. Lithuania, where important legal regulations are seen in general and in the sports sector, is very clear and open about low carbon and sustainable environmental targets in sports.

3.1 Lithuania

Lithuania, which started to change its shell after the European Union membership in 2004, is a country in central Europe neighbouring Russia, Poland and Latvia [14]. In the last 20 years, Lithuania, one of the Baltic States, has taken remarkable steps in tackling the problem of climate change. Lithuania has implemented significant economic restructuring plans, significantly reducing its emissions, especially production-based emissions [4]. Realizing that the threatening lifestyle is a threat to its sustainable world goals, Lithuania signed the Paris Agreement with its EU members, in which it pledged to reduce its greenhouse gas emissions by 40% compared to 1990. With this agreement, Lithuania has committed to reducing emissions by 9% by 2030 [31]. Based on this commitment, the Lithuanian Government has updated its sustainable development goals and set its goals under three main headings. These are environmental quality and natural resources, economic development and social development [30].

Environmental quality and protection of natural resources are at the forefront of environmental sustainability goals. Lithuania, which tries to implement the climate action plans of the European Union without any deviation, is very sensitive about protecting the environment and resources and transferring it to future generations. These responsible plans and policies have produced positive results in a short time. In fact, Lithuania's environmental performance has started to improve since the mid-2000s. CO_2 emissions have decreased significantly and are below the average of OECD countries. Waste management practices have improved significantly, with a significant reduction in waste going to landfills. One of the most important advantages of Lithuania is that forests occupy a large share of the land, about 35% of the country, and forest lands are growing. This helps offset half of the national greenhouse gas emissions. The share of these forested lands and other protected landforms within the country's borders is also increasing rapidly [22].

The European Union has determined four criteria for the solution to the global climate change problem in its climate change policy priorities. It is desired that these criteria are applied without exception and that sustainable world targets are achieved. These criteria are

1. Energy efficiency improvement or energy intensity reduction.
2. To turn to renewable energy sources and increase their share in energy consumption.
3. Reducing greenhouse gas emissions.
4. Increasing accessibility to energy [8].

In the light of the priority climate policies of the European Union, Lithuania has taken important steps throughout the country, showing that it is an important part of solving climate problems and reducing its carbon footprint. Between 2015 and 2020, energy consumption in Lithuania increased approximately 1.3 times, wind energy production increased 2.5 times, biogas energy increased 2.8 times, and total greenhouse gas emissions decreased by 3% and air pollutants by 23% [10]. It can be said that the measures taken and the orientation towards renewable energy sources are the main determinants of this decline. The Government of Lithuania has taken the issue of environmental sustainability seriously. In order to carry out this process in a healthy way, various laws, regulations, plans and policies have been implemented. At the beginning of these policies is the "Environmental Protection Law" that came into force in 1992. By making this law compatible with the European Union criteria, the Lithuanian Government has embodied its environmental sustainability goals and low carbon footprint goals.

3.2 Legal Regulations Towards Environmental Sustainability in Lithuania

3.2.1 Environmental Protection Law

With this law, which came into force in 1992, it was aimed to implement environmental sustainability plans and programs. In addition, it is aimed to protect the environment from physical, chemical, biological and other effects and results arising from economic activities or the use of natural resources. With this law, it is aimed to plan the use of natural resources within the borders of Lithuania, to protect the ecosystem, to determine environmental protection norms and to reduce the negative effects on the environment [16]. The measures to be taken for environmental sustainability and low carbon footprint in the relevant law are shaped around four titles: Four priority titles within the scope of the environmental protection law are presented visually as follows (Fig. 3.1).

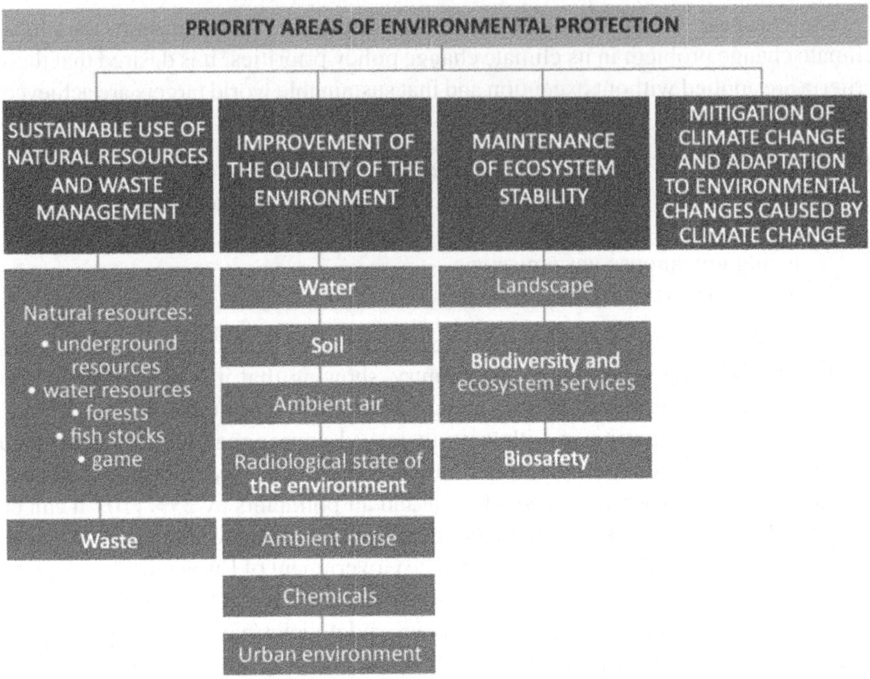

Fig. 3.1 Structure of priority areas of environmental protection in Lithuania

In order to bring the Environmental Protection Law into effect, the Ministry of Environment of the Government of Lithuania published the "Environmental Protection Strategy" document [20].

3.2.2 Environmental Protection Strategy Document

The Environmental Protection Strategy Document is accepted as a detailed and comprehensive implementation document of the implementation criteria in the Environmental Protection Law. It is aimed to protect natural resources, waste management, planned use of underground and aboveground assets, prevent air, water and soil pollution, protect bio-diversity and reduce the carbon footprint in all sectors operating throughout the country. In these areas, it points to a joint effort to reach the 2030 environmental targets, responsible and sustainable production and consumption, waste management, protection of nature, development and protection of the environment [20].

3.2.3 National Energy and Climate Action Plan

It has been prepared as a national action plan for the years 2020–2030 in accordance with the requirements specified in the Energy Union Regulation. In line with this plan, it is aimed to (a) ensure good environmental quality and harmonization of the use of natural resources, (b) reduce Lithuania's impact on climate change and increasing its resilience to its effects and (c) increase the energy efficiency of residential and public buildings [26].

3.2.4 National Climate Change Management Strategy

The "Climate Change Management" action plan put into effect by the Parliament of the Republic of Lithuania is basically an action plan created to minimize the threat of climate change. The purpose of this plan is to define Lithuania's climate change management policy by 2030, and to set goals and objectives in the areas of long-term mitigation and adaptation to the effects of climate change by 2050. Climate change mitigation policy aims to reduce the amount of greenhouse gas emissions and increase their absorption. Climate change adaptation policy aims to strengthen adaptation capacity, increase resilience and reduce vulnerability to the effects of climate change in order to contribute to sustainable development and provide appropriate counter-adaptation measures [39]. With this strategy put into practice in line with these aims, a series of principles and targets have been set in four main axes until 2030: (a) sustainable use of natural resources and waste management, (b) improving environmental quality, (c) maintaining ecosystem stability and mitigating climate change and ensuring adaptation [22].

3.2.5 National Air Pollution Reduction Plan

This action plan, which was prepared to reduce the negative effects of pollution caused by people, workplaces and facilities, on the environment and human health in all sectors operating within the borders of the Republic of Lithuania, was put into practice in 2019. The Lithuanian Government considers air pollution to be the most important human and environmental health problem. With this plan, it is aimed to control carbon dioxide and greenhouse gas emissions from all buildings in Lithuania [19].

3.2.6 Law on Construction

Building Law sets out requirements for the preparation of projects for "Accessibility of Buildings", construction, reconstruction, building renovation (modernization), capital repair, simplified construction and simplified reconstruction of structures in all sectors in Lithuania. Thanks to this law, while it is aimed that the new buildings to be built will be environmentally friendly, it is aimed to design the existing buildings to have a low carbon footprint [40]. Lithuania, which has taken important steps, especially in the renovation and improvement of buildings, has decreased energy consumption in all buildings in the country at a very serious level with the strict implementation of this law [21]. This reduction in energy consumption is due to the increasing number of refurbishments and energy efficiency improvements. In addition, depending on the liberalization of the EU energy market, the choice of renewable electricity suppliers in most buildings has also been decisive. In Lithuania, financial measures to reduce negative environmental impacts related to buildings, install solar panels, and energy efficiency measures have been implemented [17], because increasing the use of renewable energy sources provides an opportunity to develop a competitive, reliable and sustainable energy sector and to solve other important problems, for example, reducing dependency on electricity imports, increasing energy supply security, achieving targets to reduce carbon emissions and protecting the environment [1].

3.2.7 Approval of Health Strategy

The Lithuanian Health Strategy constitutes the legal framework for the process of building a healthy society covering the years 2014–2025. With this strategy document, it is aimed to reveal the aims and objectives of health promotion activities and the indicators of the level of health to be reached [38].

3.2.8 Lithuanian Building Sustainability Assessment System

The Lithuanian Green Building Council is a non-governmental organization that continues its activities throughout the country. Through this organization, the government follows sustainable building policies and practices in all sectors across the country. Thanks to the building evaluation system put forward by the council, it is aimed to reduce the carbon footprint of the buildings in the country. The Lithuanian Building Sustainability Assessment System (LBSAS) aims to objectively and transparently assess the sustainability level of districts and buildings in Lithuania and to publicly disclose the best sustainable design and construction practices [33].

Lithuanian governments have created the necessary legal basis for environmental sustainability goals, which is one of the sustainable development goals, especially after the European Union membership in 2004. In particular, it has implemented plans, policies and strategies around the planned use of natural resources in all sectors and low carbon footprint throughout the country. If it is accepted that the legal regulations throughout the country are binding for all sectors, it can be said that all the plans and policies mentioned above are also binding for the sports sector. Especially in recent years, with the negative environmental relations with the sports sector coming to the agenda intensively, Lithuania is taking significant steps in the sports sector. Sports clubs, federations and other stakeholders of sports and non-governmental organizations take the necessary steps in line with the goal of low carbon footprint in sports. The plans, policies and strategies produced for this are tried to be strictly implemented [28].

3.3 Legal Regulations Towards Environmental Sustainability in Sport in Lithuania

In order to better understand the process of scanning and examining the literature for low carbon footprint targets in the sports sector, first of all, a general review of the plans, policies and strategies prepared by the Lithuanian Governments for environmental sustainability targets has been made and an idea about the general framework of these written documents has been tried to be obtained. In this context, in the next step of the research, the environmental sustainability goals in the written documents related to the plans, policies and measures prepared primarily to reduce the carbon footprint of the sports industry and sports facilities in Lithuania and the websites of sports federations and sports clubs, which are the stakeholders of sports in Lithuania were reviewed and evaluated.

3.3.1 Law on Physical Education and Sport

"Physical Education and Sports Law" regulates the principles of physical education and sports, the powers of state and municipal institutions in the field of physical education and sports in Lithuania. Contributes to the development of physical education and sport in schools and civil society in Lithuania. It mediates the determination of requirements in areas such as sports organisation, athlete training, organisation of competitions, development of professional sport. Article 50, paragraph 2 of the relevant law, on the subject of research subject, states that "The construction, renovation and utilisation of sports facilities of national importance in Lithuania shall be promoted thorough environmental, economic and financial measures" [36].

In particular, the environmental priorities of sports facilities and the reduction of carbon footprint in sports are among the primary targets. As a matter of fact, while the new buildings to be built are aimed to be environmentally friendly in the Lithuanian Building Code, it is aimed to have a low carbon footprint in the renewal of existing buildings [40]. Barghchi et al. [2] state that the carbon footprint, which occurs due to the construction of new sports facilities, maintenance and repairs of existing ones, and the need for high energy and natural resources, is an important threat to sustainable environmental understanding.

Looking at the carbon footprint problem of sports facilities from another perspective, uncontrolled and excessive consumption of water, energy and chemicals to keep a field ready for competition causes high greenhouse gas and carbon dioxide emissions [42]. In this process, where sports facilities and environmental problems are often associated, the Lithuanian government legally guarantees the promotion of environmental practices in the construction of new sports facilities and the maintenance and repair of existing ones, while evaluating low carbon footprint targets in a legal framework. In addition, high amounts of energy are needed for the active and effective operation of sports facilities, along with the increased use of air conditioning, water and electricity due to heating, cooling and lighting of the host facilities during sports organizations and events [15]. However, as a result of the consumption behaviour of the participants during sporting events, it causes large amounts of waste to be generated, which contributes to the carbon footprint [5].

3.3.2 *Law on Charity and Sponsorship*

With this law, it is assumed that sports events and activities are of public interest. It is based on the protection of the relevant rights and the environment in and around the sports facilities where sports activities are carried out [35]. For sustainable sports events and organizations, rehabilitating the facilities and the environment of the facilities with an environmentally friendly approach is decisive in reducing the carbon footprint. Trendafilova et al. [41] state that the sports sector is aware of the environmental problems in question and that it has set two goals within itself. These are to reduce the ecological footprint of sports and to increase the awareness of the participants about the importance of protecting the environment. For example, mass participation in organized sports events and organizations is the primary cause of carbon footprint. These participations, called sports tourism, are the travels of the athletes and spectators to participate in and follow the competitions. Sports tourism is defined as leisure-based travel that takes individuals temporarily out of their country or city to participate in, watch sports organizations or reach attractions related to organizations [11]. Therefore, these travels cause the formation and growth of the carbon footprint.

3.3.3 Law on Public Organizations

With this law, which has been in force since 1998, one of the important criteria in the use of the funds provided for sports events and organizations is to take environmental protection and other measures stipulated by legal regulations [37]. In addition to the measures related to sports facilities in Lithuania, the necessity of using the funds provided for the planning and implementation of public sports events also arises for environmental protection plans. Sustainable environment and sports targets overlap with this article of law. Facilities hosting amateur or professional sporting events have significant adverse environmental impacts. Sorrentini [29] states that mass participation in sports activities causes transportation, purchasing, accommodation, purchasing and an intensive waste production depending on all these, and this situation feeds the carbon footprint problem in sports.

3.3.4 Green Coach Project

GREEN COACH—"Integrating sustainability into the management of sports facilities and organizations" is a project that aims to improve good governance in terms of environmental impacts in sports organizations by monitoring their environmental footprints and adopting sustainable behaviours in their daily management strategies. This project was funded by the European Union. And 5 sports clubs, especially the Lithuanian Football Federation, were involved in the project. Within the scope of this project, a roadmap was created to reduce the carbon footprint in the sports industry. Lithuanian Football Federation stated that other sports clubs will fulfil the commitments of the project. The action plans prepared for reducing the carbon footprint in sports facilities and organizations within the scope of the project are as follows:

- Classifying and collecting waste separately in sports facilities.
- Meeting the energy needs of sports facilities from renewable energy sources.
- Increasing the use of LED lamps to illuminate the facilities.
- Controlling water usage in and around the facilities and installing water flow reducers for this.
- Promoting the use of recyclable materials in facilities and organizations.
- Increasing the use of disposable or reusable beverage bottles [7].

Lithuanian Football Federation and 5 other sports clubs have committed to implementing these criteria determined as a result of the project in sports facilities and events and stated that steps will be taken to reduce the carbon footprint in sports facilities [7].

The classification of waste and the establishment of related solid waste management is a priority to achieve the European Union's sustainable development goal, which is to reduce the negative impact on the environment and human health [24]. One of the most important steps regarding waste management in Lithuania was the introduction of the deposit system [32]. Thanks to the deposit system, 92% of all packaging (glass, plastic or aluminium) launched in 2018 was returned for recycling. Lithuania has one of the highest recycling growth rates among European countries over a 15-year period [9]. Another reform was the introduction of an electronic system to monitor the movement of waste from generation to transport and disposal [24]. This situation has been put into practice in all sectors, including the sports sector in Lithuania, and economic and environmental positive outputs have begun to be obtained in a short time.

Lithuania, like other European countries with cold climates, has well-developed district heating systems. Lithuania's national energy strategy includes reducing dependency on imported fossil fuels, reducing the amount of greenhouse gases released into the atmosphere, and using sustainable renewable energy sources throughout the country [13], because, considering that fossil fuels play the most important role in providing the energy required for the industrial process, it is inevitable to destroy the environment and release large amounts of carbon, wastewater and waste gas as undesirable outputs [44, 45].

Sports facilities are one of the buildings that causes high energy consumption. The fact that this energy need is met from fossil fuel sources is one of the most important causes of the carbon footprint in sports. However, today, renewable energy sources surpass fossil fuels as the largest energy source and become the core of achieving sustainable climate goals on a global scale [12]. Constructing sports facilities that are sensitive to nature and the environment, having low energy consumption within these facilities, facilitating access to existing sports facilities, disseminating environmentally friendly practices such as bicycle use and ensuring the continuity of sports activities as well as a sustainable environment can contribute to individual and social health and the living environment [23].

3.3.5 *The Lithuanian National Olympic Committee*

The Lithuanian National Olympic Committee (LNOC) is an autonomous, nonprofit public organization belonging to the global Olympic Movement. LNOC conducts its activities in accordance with the Constitution and legislation of the Republic of Lithuania, the Olympic Statutes, the World Anti-Doping Code, International Olympic Committee (IOC) resolutions and LNOC regulations. The Olympic committee expresses its responsibility for environmental sustainability on its website under the heading of goals, objectives and strategies as follows: "acting for peace, promoting and developing sports ethics, combating doping and dealing with environmental problems responsibly". Although the National Olympic Committee does not make a comprehensive description of its environmental responsibilities, it is committed to complying with the decisions and regulations of the International Olympic Committee. Therefore, it is understood that the International Olympic Committee has implemented the "Sustainability Strategies" [34].

3.3.6 International Olympic Committee Sustainability Strategy

The International Olympic Committee has published the report titled "Sustainability Strategies", which it has decided to implement in all member countries. Lithuania, one of the member countries, is trying to implement the titles in this strategy document in sports facilities within the borders of the country. In this strategy document, the International Olympic Committee has targeted the management of direct and indirect greenhouse gas emissions associated with the activities of the Olympic Movement and the adaptation of sports facilities and events to the consequences of climate change. The importance of environmental sustainability is highlighted in the relevant strategy document as follows:

> We are aware of the need for urgent action to protect the climate and adapt to the changing climate. Various activities within the Olympic Movement generate greenhouse gas emissions directly (e.g. emissions from IOC vehicles and buildings) or indirectly (e.g. air travel, goods and services purchased) and these emissions need to be further reduced.
>
> Assessing the long-term viability of infrastructure built for the Olympic Games in host countries is an important part of sustainable environmental goals. Many cities in sports facilities around the world already have climate mitigation and adaptation plans in place; in this context, hosting the Olympic Games should be seen as a way to support and accelerate the realization of these plans. Because host countries and cities play a key role in addressing the global climate change challenge [12].

In order for any country or city to host an event, the International Olympic Committee is expected to fulfil the following criteria regarding environmental sustainability and contribute to the reduction of carbon footprint in sports:

- Conservation of biodiversity.
- Protection of the ecosystem.
- Correct land use and landscaping studies.
- Prevention of pollution (soil, air, water).
- Resource and waste management.
- Health and safety measures.
- Reducing environmental disturbances.
- Protection of cultural heritage [12].

Lithuanian National Olympic Committee, which is one of the member countries of the International Olympic Committee, is obliged to take the above-mentioned measures in order to reduce the carbon footprint in sports facilities and organizations within the borders of the country. It aims to minimize the negative environmental effects of the sports sector, especially by reducing the carbon footprint due to high energy consumption and creating waste management systems.

3.3.7 Addressing Climate Change Through Sport

It is a strategy document prepared by the United Nations, Department of Economic and Social Affairs. It is expected that the strategy document, which draws attention to the need to produce solutions to climate change through sports, will be accepted and implemented as a reference by all member countries. Lithuania, a member of the United Nations, has included this strategy document in its climate action plans through sports. As a matter of fact, in this strategy document, the action plans of national governments are frequently emphasized [43]. The measures to be taken in this strategy document, which draws attention to the fact that the carbon footprint can be reduced in sports with the action plans of national governments, are listed as follows:

- Member states' sports organizations should commit to reducing their direct and indirect greenhouse gas emissions and take concrete steps (using renewable energy, building carbon-free gyms, etc.);
- Governments should set policy frameworks and incentives to make sport sustainable and to promote sport as a means of raising awareness of climate change;
- Undertake capacity-building efforts targeting different actors in the sports sector on how to address and mitigate the adverse environmental impacts of the sports sector, in collaboration with national governments, international organizations and other national actors;
- As measuring and reporting greenhouse gas emissions are important steps towards the implementation of the Paris Agreement, national governments should collect robust data in collaboration with international organizations. Estimates need to go beyond current narrow-focus assessments and include the broader range of carbon emissions associated with sporting events;
- National governments and international organizations should create collaborations and partnerships between different stakeholders, including national and local governments, the private sector, academia, science, technology and innovation experts, and non-governmental organizations to promote sport as a climate action tool [43].

Since hosting sports events is important for many different reasons for countries today, environmental effects have become particularly evident when it comes to carbon footprints, and the sports sector has become an important part of the climate change problem [18]. The Lithuanian National Olympic Committee and sports federations, which have implemented measures to reduce the carbon footprint in many different branches, especially football and basketball, try to take an active part in the climate action plans of the United Nations, trying to take the necessary steps in different sports. Lithuania works for environmental sustainability targets. One of the most striking examples of these works is Zalgiris, one of the most prominent sports clubs in the country. It is Kaunas. The club has implemented the Environmental Protection Program in the light of the sustainable environmental goals of international sports organizations and national environmental protection policies.

Zalgiris, one of the most important clubs of Lithuanian and European basketball, plays its competitions at the Zalgirio Arena in Kaunas. This facility, which stands out, especially with its environmentally friendly plans, is operated in the light of the Environmental Protection Policies put into practice by the Lithuanian government. According to this policy, one of the established green policies of the Zalgirio arena is to aim to reduce the negative impact of the Arena and its ongoing activities on nature, the environment, health and people in general [46]. In the light of these purposes, Zalgiris Kaunas club has set the following goals:

- **Environmental Protection Policies of the Basketball Hall**:
  - In 2014, an agreement was signed to initiate an environmental policy that aims to reduce environmental risks and includes recommendations for gym users, employees and guests.
  - Fans who come to the hall for competitions are encouraged to recycle garbage by separating food waste from paper, plastic and glass. The garbage separation process is very important. For example, plastic is a non-renewable material that does not degrade in the environment. The materials used in the production of plastics are cellulose, coal, natural gas, salt and crude oil. Therefore, plastic directly resembles the negative environmental effects of fossil fuels. Therefore, the negative environmental effects of the facilities can be minimized by collecting these wastes in different garbage containers according to their types [27].
  - In order to ensure correct recycling, all garbage in the hall is rechecked and recycled if necessary.
  - In the light of all these targets, it was aimed to reduce the waste originating from the gym by 5% until 2017 in general, and this aim was achieved.
- **Use of Organic Materials**:
  - Žalgirio arena only works with suppliers that work with ISO 14001 or similar environmental protection systems. The cleaning materials used are environmentally friendly and "European Ecolabe" or have documents proving their contents.
- **Reducing Vehicle Emissions**:
  - To ensure lower emissions, arena staff and visitors are encouraged to reach the arena by bicycle or public transport.
  - There are bike carriers specially made to accommodate green transportation, and there are also special buses with the "ZA" (electric) sign that take the arena visitors to different parts of the city after the big events in the arena [46].

In line with the above-mentioned objectives, Zalgiris Kaunas Basketball Club gives priority to reducing its carbon footprint in all training and competition facilities, competitions and other events. It plans its future in the light of the sustainable environment and sports goals by emphasizing the environmentally friendly sports and facility management approach.

Reducing the carbon footprint in sports at the level of sports federations and government bodies is one of the priorities of the Lithuanian sports management approach. In Lithuania, for example, there are many event organizers who do not give up on disposable plastic items at sporting events. The Lithuanian capital council took an extraordinary decision and stated that disposable plastic containers will not be used in events that will take place in Vilnius from 2020. This solution will help reduce the amount of plastic waste in the city and educate citizens on waste management [28].

Lithuania, which has a deep-rooted historical background, joined the European Union after 2004 and perhaps started the modernization process. One of the most important parts of this process is sportive development. However, together with sportive development, the protection of its natural and environmental wealth has also been a priority issue.

As in other countries, there are intense travels in Lithuania due to the construction of the sports sector and participation in organizations. This situation carries risks to the environment. In order to minimize these risks, the Lithuanian Governments have prepared the legal ground and continue to work in cooperation with international sports organizations. While meeting the increasing sportive demand, it is also aimed to protect the natural wealth of the country. It is also aimed to design legal regulations in accordance with the era and to further international cooperation.

References

1. T. Baležentis, D. Štreimikienė, R. Melnikienė, S. Zeng, Prospects of green growth in the electricity sector in Baltic States: pinch analysis based on ecological footprint. Resour. Conserv. Recycl. **142**, 37–48 (2019). https://doi.org/10.1016/j.resconrec.2018.11.013
2. M. Barghchi, D. Omar, M.S. Aman, Sports facilities in urban areas: trends and development considerations. Pertanika J. Soc. Sci. Humanit. **18**(2), 427–435 (2010)
3. J. Brizga, I. Kudreņickis, Household climate impact in Latvia: measuring carbon footprint. Environ. Clim. Technol. **3**(3), 34–40 (2009)
4. J. Brizga, K. Feng, K. Hubacek, Drivers of greenhouse gas emissions in the Baltic States: a structural decomposition analysis. Ecol. Econ. **98**, 22–28 (2014). https://doi.org/10.1016/j.ecolecon.2013.12.001
5. J.M. Casper, M.E. Pfahl, Environmental sustainability practices in US NCAA Division III athletics departments. Int. J. Event Manag. Res. **10** (2015)
6. J. Chen, W. Fan, D. Li, X. Liu, M. Song, Driving factors of global carbon footprint pressure: based on vegetation carbon sequestration. Appl. Energy **267**, 114–127 (2020). https://doi.org/10.1016/j.apenergy.2020.114914
7. T. Daddi, L. Marrucci, R. Stranieri, D. Casiddu, Evidence based study and action plan to decrease the environmental footprint of grassroots sports: intellectual output 3 sustainable football. EU Green Coach Project Reports (2022)

8. European Commission, National Renewable Energy Action Plans—EU Science Hub—European Commission (2020). https://ec.europa.eu/energy/en/topics/renewable-energy/national-renewable-energy-action-plans-2020v. Accessed 17 Mar. 2023
9. European Environment Agency, Municipal waste recycling rates in Europe by Country (2019). https://www.eea.europa.eu/ims/waste-recycling-in-europe. Accessed 27 Mar. 2023
10. V. Gaigalis, V. Katinas, Analysis of the renewable energy implementation and prediction prospects in compliance with the EU policy: a case of Lithuania. Renew. Energy **151**:1016–1027. https://doi.org/10.1016/j.renene.2019.11.091
11. H.J. Gibson, Sport tourism: a critical analysis of research. Sport Manag. Rev. **1**(1), 45–76. https://doi.org/10.1016/S1441-3523(98)70099-3
12. International Energy Agency, World Energy Outlook 2016 (2016). https://www.iea.org/reports/world-energy-outlook-2016. Accessed 17 Mar. 2023; International Olimpic Comitte, Sustainability report (2021). https://stillmed.olympics.com/media/Documents/News/2021/12/IOC-Sustainability-Report-2021.pdf?_ga=2.25100083.827008892.1671631206-1564112537.1670779546
13. R. Jonynas, E. Puida, R. Poškas, L. Paukštaitis, H. Jouhara, J. Gudzinskas, G. Miliauskas, V. Lukoševičius, Renewables for district heating: the case of Lithuania. Energy **211**(2020), 1–11 (2020). https://doi.org/10.1016/j.energy.2020.119064
14. T. Karpavicius, T. Balezentis, Public service obligation levy in the context of energy sustainability and security: the cases of Ireland, Greece, Denmark and Lithuania. Energies **15**(1), 16–27 (2021). https://doi.org/10.3390/en15010016
15. T.B. Kellison, M.J. Mondello, Organizational perception management in sport: the use of corporate pro-environmental behaviour for desired facility referenda outcomes. Sport Manag. Rev. **15**(4), 500–512 (2012). https://doi.org/10.1016/j.smr.2012.01.005
16. Law on Environmental Protection (1992). https://e-seimas.lrs.lt/portal/legalAct/lt/TAD/6378f2b0023211e6bf4ee4a6d3cdb874. Accessed 19 Mar. 2023
17. G. Liobikienė, J. Brizga, Sustainable consumption in the Baltic States: the carbon footprint in the household sector. Sustainability **14**(3), 1–14 (2022). https://doi.org/10.3390/su14031567
18. S. Manfredi, D. Tonini, T.H. Christensen, H. Scharff, Landfilling of waste: accounting of greenhouse gases and global warming contributions. Waste Manag. Res. **27**, 825–836 (2009)
19. Ministry of Environment of the Republic of Lithuania, National Air Pollution Reduction Plan (2019). https://e-seimas.lrs.lt/portal/legalAct/lt/TAD/6378f2b0023211e6bf4ee4a6d3cdb874. Accessed 17 Mar. 2023
20. National Environmental Protection Strategy (2016). https://am.lrv.lt/uploads/am/documents/files/National%20Environmental%20Protection%20Strategy.pdf. Accessed 17 Mar. 2023
21. Odyssee-Mure, Energy efficiency indicators for Europe-key indicators. Enerdata (2021). https://www.odyssee-mure.eu/. Accessed 17 Mar. 2023
22. Organisation for Economic Co-operation and Development, Greening Lithuania's Growth Economics Department Working Papers No. 1667 (2021). https://www.oecd.org/officialdocuments/publicdisplaydocumentpdf/?cote=ECO/WKP(2021)18&docLanguage=En. Accessed 27 Mar. 2023
23. C.A. Öztekin, Sürdürülebilir yaşam rehberi. J. Manag. Theory Pract. Res. **2**(2), 178–181 (2020)
24. A. Paulauskaite-Taraseviciene, V. Raudonis, K. Sutiene, Forecasting municipal solid waste in Lithuania by incorporating socioeconomic and geographical factors. Waste Manag. **140**, 31–39 (2022). https://doi.org/10.1016/j.wasman.2022.01.004
25. A. Radu, M. Scrieciu, D. Carocota, Carbon footprint analysis: towards a projects evaluation model for promoting sustainable development. Procedia Econ. Financ. **6**(2013), 353–363 (2013). https://doi.org/10.1016/S2212-5671(13)00149-4
26. Republic of Lithuania, National Energy and Climate Action Plan (2021a). https://am.lrv.lt/uploads/am/documents/files/KLIMATO%20KAITA/Integruotas%20planas/Final%20NECP.pdf. Accessed 27 Mar. 2023
27. K. Sabaliauskaitė, D. Kliaugaitė, Resource efficiency and carbon footprint minimization in manufacture of plastic products. Environ. Res. Eng. Manag. **1**(67), 25–34 (2014). https://doi.org/10.5755/j01.erem.67.1.6587

28. S. Skirmantas, B. Svagzdiene, Sustainability evaluation of sports events organized in Lithuania. Humanit. Stud. **5**(82), 160–171 (2020). https://doi.org/10.26661/hst-2020-5-82-12
29. F. Sorrentini, The environmental impact of sports activities. Good practices for sustainability: the case of golf. Documenti Geografici **2**, 219–237 (2021). https://doi.org/10.19246/DOCUGEO2281-7549/202102_15
30. E. Štuopyte, Towards sustainability: The involvement of the elderly in the educational activities of NGOs in Lithuania. Sustainability **14**(4), 2–12 (2022). https://doi.org/10.3390/su14042095
31. Sustainable Governance Indicators, Governance, Sustainable Policies, Democracy Lithuanian Report (2022). https://www.sgi-network.org/docs/2022/country/SGI2022_Lithuania.pdf. Accessed 17 Mar. 2023
32. G. Tamulyte, Waste Management in Lithuania (2021).https://wilderness-society.org/waste-management-in-lithuania. Accessed 17 Mar. 2023
33. The Lithuania Green Building Council, Lithuanian building sustainability assessment system (2013). https://www.lzpt.lt/about-us/. Accessed 17 Mar. 2023
34. The Lithuanian National Olympic Committee, About The Lithuanian National Olympic Committee (2022). https://www.ltok.lt/en/about-lnoc/. Accessed 18 Mar. 2023
35. The Republic of Lithuania, Law on Charity and Sponsorship (1993). https://e-seimas.lrs.lt/portal/legalAct/lt/TAD/TAIS.281044?jfwid=uu1o99phi. Accessed 19 Mar. 2023
36. The Republic of Lithuania, Law on Physical Education and Sport (1995). https://e-seimas.lrs.lt/portal/legalAct/lt/TAD/a69875e0e9da11e59b76f36d7fa634f8?jfwid=2r1mca2o. Accessed 20 Mar. 2023
37. The Republic of Lithuania, Law on Public Organization (1998). https://e-seimas.lrs.lt/rs/legalact/TAD/TAIS.71102/. Accessed 20 Mar. 2023
38. The Republic of Lithuania, National, Approval of Health Strategy (2014). https://e-seimas.lrs.lt/rs/legalact/TAD/608a896236f811e6a222b0cd86c2adfc/. Accessed 20 Mar. 2023
39. The Republic of Lithuania, National Climate Change Management (2021b). https://unfccc.int/documents/461231. Accessed 20 Mar. 2023
40. The Republic of Lithuania, Law on Constriction (1996). https://e-seimas.lrs.lt/portal/legalAct/lt/TAD/TAIS.312477?jfwid=bkaxmm6x. Accessed 20 Mar. 2023
41. S. Trendafilova, B. McCullough, M. Pfahl, S.N. Nguyen, J. Casper, M. Picariello, Environmental sustainability in sport: current state and future trends. Glob. J. Adv. Pure Appl. Sci. **3**, 9–14 (2014)
42. United Nations Climate Change, Creating the greenest football club in the World-Forest Green Rovers (2022). https://unfccc.int/climate-action/sectoral-engagement/sports-for-climate-action. Accessed 21 Mar. 2023
43. United Union, Department of Economic and Social Affairs, Addressing Climate Change through Sport (2022). https://www.un-ilibrary.org/content/papers/https://doi.org/10.18356/27081990-128. Accessed 22 Mar. 2023
44. J. Wu, P. Yin, J. Sun, J. Chu, L. Liang, Evaluating the environmental efficiency of a two-stage system with undesired outputs by a DEA approach: an interest preference perspective. Eur. J. Oper. Res. **254**(3), 1047–1062 (2016). https://doi.org/10.1016/j.ejor.2016.04.034
45. T. Xu, J. You, H. Li, L. Shao, Energy efficiency evaluation based on data envelopment analysis: a literature review. Energies **13**(14), 35–48 (2020). https://doi.org/10.3390/en13143548
46. Zalgiris Kaunas Environmental Protection Programme, Environmental protection (2014). https://zalgirioarena.lt/en/aplinkosauga/. Accessed 21 Mar. 2023

Chapter 4
Official Guide to Mitigate the Negative Effects on Nature in Sport

This part of the study includes suggestions that stakeholders within the sports sector can benefit from. In the editing and compilation of these recommendations, "***EU Project that A Roadmap for Creating a Sustainability Strategy***" was used [22].[1]

First of all, the importance of environmental sustainability for the sports sector was emphasized, and then, institutional suggestions were presented so that sports institutions could exhibit an environmentally friendly understanding. Then, suggestions for environmentally friendly management for sports facilities and events are given. These suggestions can be stretched according to the influence of the stakeholders of the sport, or better results can be targeted by developing them. However, the implementation of the suggestions presented at the basic level with realistic practices is very important in terms of the protection of the natural environment and the sustainability of sports.

4.1 First Step: The Importance of Environmental Sustainability for the Sports Sector

The sports sector, with all its components, is one of the areas where human-based activities are most intense. On a global scale, the intensity of active and passive participation in all branches, especially football, is increasing day by day. Increasing Interest and demand are growing in the sports industry day by day. Every growth can bring some problems with it. One of the problems that has been frequently

[1] *The Roadmap for Creating a Sustainability Strategy was developed within the framework of the "As Sustainable As Possible (ASAP) project" co-funded by the Erasmus + Programme of the European Union. The European Commission's support for the production of this publication does not constitute an endorsement of its contents, which reflect the views only of the authors, and the Commission cannot be held responsible for any use which may be made of the information contained therein.*

A. Atalay and B. Švagždienė, *Achieving Environmental Sustainability in the Sports Sector*, SpringerBriefs in Applied Sciences and Technology,
https://doi.org/10.1007/978-3-031-37401-2_4

discussed in the sports industry in recent years and caused by growth is the negative environmental effects caused by the sports industry. Until the last decade, it is not possible to say that the handling and evaluation of negative environmental effects within the sports industry have been very efficient when compared to other sectors. Therefore, it can be said that the action plans against negative environmental effects are insufficient and are overshadowed by economic concerns and sportive competition. However, with the recent steps taken by international sports organizations (IOC, UEFA, FIFA), it can be stated that environmental problems have begun to find more place in the sports sector. In particular, in the last world cup held in Qatar 2022, sports and environmental sustainability were handled together, and it was seen that environmentally friendly practices were implemented in addition to commercial and sportive plans for this world cup.

Aiming at an environmentally friendly management approach in sports, this *roadmap focuses on a carbon-friendly sports management approach in the sports industry. The most important part of this understanding is sports organizations and facilities that have a low carbon footprint. As a matter of fact, one of the sub-definitions of the climate change problem today is the carbon footprint. Carbon footprint can be expressed as greenhouse gas and carbon dioxide emissions caused by human-based activities. With the increase in human-based activities, the carbon footprint is growing day by day and poses a threat to the sustainable environment.*

The sports sector is one of the areas where human-based activities are concentrated. The number of events and organizations organized in line with the increasing demand in sports is increasing. It is seen that the establishment of sports in the execution of these activities and organizations has gained momentum. The embodiment of the negative environmental effects of these facilities and buildings, which are seen as a solution to the infrastructure problem in construction and sports, can be called the "Carbon Footprint". Therefore, a roadmap is needed in order to reduce the negative environmental effects of sports organizations and facilities and to reduce the carbon footprint.

A roadmap has been prepared for all stakeholders of sports with the aim of implementing an environmentally friendly management approach in sports management. This roadmap consists of two steps:

- Sports facilities with a low carbon footprint: Suggestions have been made to reduce the carbon footprint resulting from events, tournaments and activities held in sports facilities.
- Organizations with a low carbon footprint: Suggestions have been presented to reduce the carbon footprint resulting from participation in all small, medium and large-scale sports organizations.

Sports organizations and facilities are the scene of intense human activities due to mass participation. As presented in the diagram below, the carbon footprint in the sports sector arises for four different reasons. These are high energy consumption, travel, water consumption and waste generation.

Energy consumption arises for the heating, cooling and lighting needs of sports facilities. However, meeting this energy need from fossil fuel sources (coal, natural gas, etc.) is considered an environmental threat. There is continuity in amateur or professional sports. In this case, there is a constant need for lighting, heating and cooling of facilities, stadiums and pools. This situation is an important threat to the environment and nature due to carbon and greenhouse gas emissions. Especially when thousands of people come together for a competition, it triggers high energy consumption. This is an important issue that needs to be addressed for environmental sustainability. In addition to carbon and greenhouse gas emissions, sports facilities and activities have devastating effects on the natural environment, ecosystem and habitat. Air, soil and water pollution is a great risk to the natural balance. Especially the wastes produced due to sports activities can be considered as sources of air, soil and water pollution.

In addition to the consumption of energy and natural resources in the sports sector, another source of environmental threat is transportation. As stated in several different points of this study, the busy sports calendar causes teams and fans to travel constantly. Today, it is known that thousands of people are on the move in many popular sports branches, especially football, and they travel nationally and internationally to watch the competitions. Travel in the sports industry can seem like an endless loop because every week, the fans travel all over the country for the competitions played in the local leagues. Carbon and greenhouse gas emissions caused by the fuels of the transportation vehicles used in these travels accelerate environmental pollution and deterioration.

Another problem associated with the sports sector is the unplanned use of natural resources. At the beginning of these resources are clean water resources. However, it can be said that clean water resources are used unplanned in facilities and organizations for different purposes. It is known that clean water resources are also in danger, especially with the threat of global warming. Finally, it is the intensive production of waste caused by mass participation events. Uncertainty of processes related to tracking, collection and disposal of waste is an important part of environmental problems. In general, the carbon footprint sources emerging in the sports sector are presented by visualizing as follows (Fig. 4.1).

Reducing the carbon footprint can be achieved at the end of a certain managerial process. This process needs to be carried out and evaluated effectively. A healthy and effective management model should be implemented, especially for an environmentally friendly management approach. There are certain steps that must be followed in the operation of this management model.

First of all, it is necessary to obtain data on carbon footprint sources caused by sports facilities or organized organizations and these data should be consistent. Then, it is inevitable to pay attention to the comparability, transparency and accessibility of these data. Then, the consumption habits in the events, organizations and tournaments and sports facilities should be controlled. In the next step, it is essential that these consumption habits are environmentally friendly and that these consumption habits are supported with environmentally friendly products. Taking measures to reduce the carbon footprint depending on the amount of energy consumed, waste and garbage produced can be considered as the last step of this process. The management approach to be created to reduce the carbon footprint is visualized as follows (Fig. 4.2).

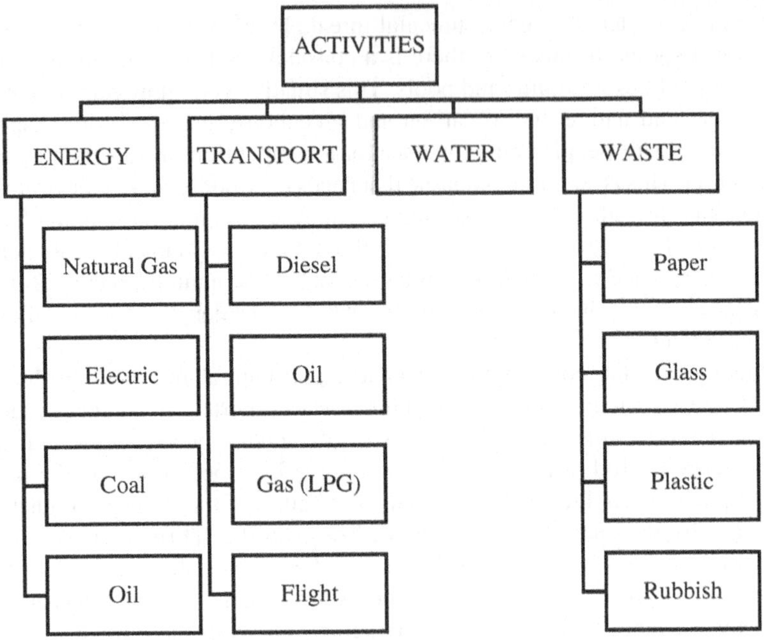

Fig. 4.1 Resources of carbon footprint in sport (Image created by researcher)

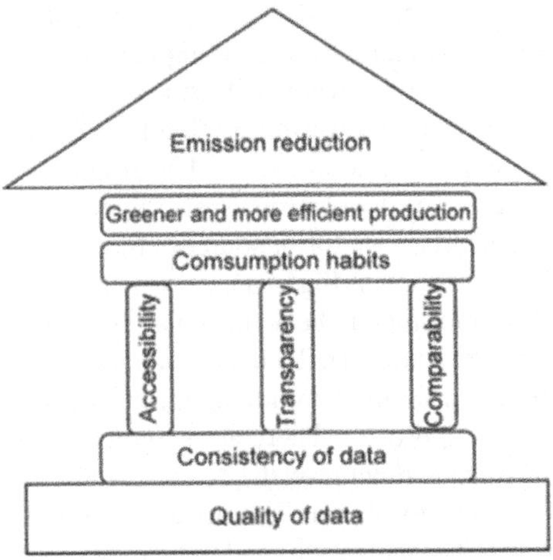

Fig. 4.2 Requirements system for reducing carbon footprint. *Source* Villar et al. [25]

Calculating the carbon footprint in the sports sector and reducing this footprint are possible by integrating an environmentally friendly management approach into sports management. By obtaining valid and healthy data and calculating these data in a transparent and accountable way in order to make comprehensive evaluations, carbon footprint calculations of facilities and organizations can be made, and then practices in line with environmental sustainability targets can be implemented to reduce this permit. Protecting the economic value of the sports sector and establishing environmentally friendly facility and organization management principles are of vital importance for both increasing its financial value and realizing sustainable sports and environmental goals. The sports sector, which is an important part of combating climate change and sustainable environmental activities, has the potential to play a very important role in transferring the environment we live in to future generations. In addition, the motto of sustainable sports can be expanded and it can be aimed to build an environmentally friendly, economical and sportive competition.

Why Should the Sports Industry Have a Low carbon Footprint ?

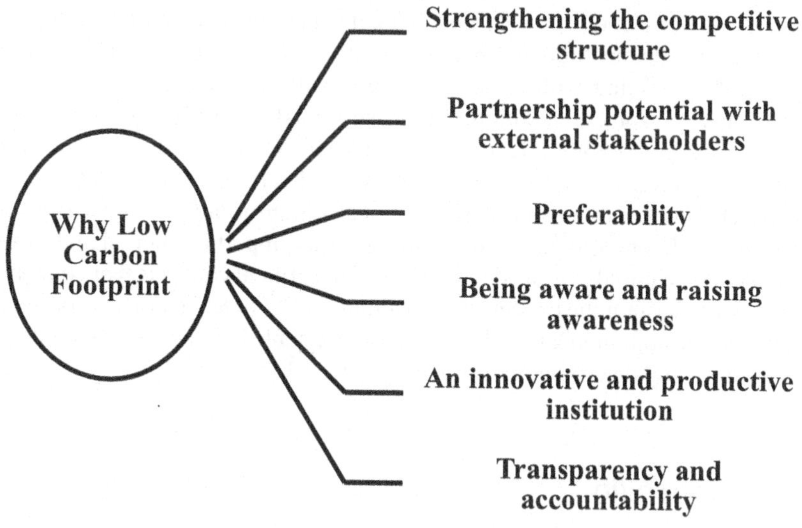

The sustainable environment goal is part of the fight against global climate change. However, in the sports sector, institutions need to be a part of this struggle and strengthen their own structures in this process because, with the increasing competition in the sports sector, the importance of strong institutional structures is increasing day by day, as in other sectors. While sports organizations with low carbon footprints are taking steps towards sustainable sports targets in competition, it can also assume an important responsibility for the sustainable environment.

4.1.1 Strengthening the Competitive Structure

Institutions, establishments, clubs and organizations in the sports sector can strengthen their corporate image in the sector by developing environmental plans and policies, thus gaining an advantageous position in intense financial and sportive competition. They can improve their ability to compete. In terms of strengthening the competitive structure of sports organizations and preparing a sustainable environment [21], it is essential to consider environmental sustainability policies and understandings.

4.1.2 Partnership Potential with External Stakeholders

Partnerships can be established with national and international institutions as a solution partner for the relationship between the academy and the sports sector, which is getting stronger in theory. As an important part of social and environmentally friendly campaigns, corporate image and brand value can be increased. Today, it has been understood that institutions are not institutions that produce goods and provide services only by acting within the framework of their economic responsibilities, and thus make a profit. In addition, they have very different functions in terms of making the world more liveable, developing society and ensuring the sustainability of limited resources. Institutions and organizations in all sectors have to adapt to the changing era and conditions brought by globalization in order to gain a competitive advantage. In order to ensure this harmony, the support of internal and external stakeholders is required [4]. In addition, one of the critical concepts that come to the fore in order to sustain the existence of institutions and organizations is stakeholder support, and the communication that institutions establish with their stakeholders is the most important determinant of reputation levels [26].

4.1.3 Preferability

Sports is a sector demanded by all segments of society. Institutions, establishments, clubs and organizations in the sector can create an environmentally friendly approach and low carbon footprint policies, increase awareness of environmental problems, leave a positive impact on target audiences and increase preference. Reputation is one of the most important values that institutions and organizations have in providing a competitive advantage in the sector. A strong corporate reputation can increase public trust, reduce stakeholder uncertainty, increase competitiveness and isolate the organization in times of crisis or threat [8]. While a strong reputation image is decisive in the preference of institutions, it can strengthen the corporate structure.

4.1.4 Being Aware and Raising Awareness

To act with awareness about the negative environmental effects caused by the sports sector, especially climate change. To raise awareness of target audiences by drawing attention to these problems in the society. One of the most important social responsibilities of today's businesses is activities aimed at protecting the environment [2]. Behaving proactively in environmental problems, developing inclusive measures, creating environmental volunteerism and awareness, supporting environmental programs, including environmental elements in the vision, mission and activities of the enterprise and providing trainings to increase environmental awareness for employees and the society [1].

Until recent years, it is not possible to say that the handling and evaluation of negative environmental effects in the sports industry is very efficient when compared to other sectors. Therefore, it can be said that the action plans created against environmental effects are weak and overshadowed by economic concerns and sportive competition. However, with the recent steps taken by international sports organizations (IOC, UEFA, FIFA), it can be stated that environmental problems have begun to find more place in the sports sector. In particular, in the last world cup held in Qatar 2022, sports and environmental sustainability were handled together, and it was seen that environmentally friendly practices were also implemented in addition to commercial and sportive plans for this world cup [5].

4.1.5 An Innovative and Productive Institution

As a solution partner for environmental problems, producing innovative solutions to reduce the carbon footprint, developing solution partnerships, taking responsibility; take productive and concrete steps. To create an institutional structure that initiates the change in the solution of environmental problems and can keep up with the changes. In this way, it is possible to produce new products and services, thus revealing productive processes in organizations [15].

4.1.6 A Transparent and Accountable Institution

To create a transparent corporate image about environmental policies and financial, social and environmental outputs of intangible and tangible inputs in production and consumption processes. To be accountable to both the relevant institutions and the groups in the society for which it is responsible, thanks to its transparent structure. To create a transparent and accountable corporate structure focused on value creation.

In terms of country image, a separate paragraph should be opened to the question of why sports management targets a low carbon footprint because international sports organizations such as IOC, UEFA and FIFA evaluate the environmental sustainability policies of the countries that will host international sports organizations, and these assessments appear as an important criterion.

Increasing the possibility of hosting organizations that are nominated due to their political, economic and social benefits, especially in the international arena, can form the basis of the understanding of "sports management aiming at low carbon footprint". The fact that countries act with a low carbon footprint facility model and an environmentally friendly sports management approach, and the spread of an environmentally friendly approach to the whole country can "increase the potential of hosting international competitions". In this section, where we try to reveal the general framework of the relationship between sports and environmental sustainability, suggestions for sports facilities, organizations and activities are presented.

4.2 Second Step: Institutional Recommendations

The International Olympic Committee recognizes sport as an important provider of sustainable development. Meeting the demands and expectations of the societies on a global scale is possible with the understanding of corporate governance. One of the corporate management trends today is the harmony of sports facilities and organizations with the natural environment. When sustainable sports can be possible with a sustainable environment, both the literature and practitioners should handle this process in a more scientific and professional way.

Sports management in general and the management setup of sports facilities in particular should be designed with a corporate management approach. Specifically, it should be ensured that the understanding of environmental sustainability is one of the most important steps of the corporate governance model. While planning the managerial process, the creation, implementation and control of environmental sustainability strategies express the generality of this process. Unlike other sectors, sports organizations have a social and lively structure. Their agendas are intense and variable. It has a structure that has no difficulty in keeping up with the rapidly changing social structures in today's world. Environmental sustainability is a very popular and frequently mentioned issue. Especially with the global climate problem, the increase in temperatures and the very rapid consumption of natural resources are an issue that concerns all sectors as well as the sports sector. Therefore, the environmental issue should be handled professionally in the institutional structures and management processes of sports organizations.

So, what steps should be taken and strategies should be implemented in the corporate governance model in the sports sector for environmental sustainability within the institution and with other institutions that the institution is in contact with? In order to answer these questions, a number of institutional recommendations have been presented.

- First of all, they need to create a policy document that clubs or sports organizations should follow. This document should primarily contain a situation analysis. In other words, what is the status of the sports institution regarding environmental problems? What are they doing? It should have a fiction that analyses the current situation. This policy document should contain concrete and measurable goals and objectives. Being realistic and traceable can be a source of motivation for sports institutions in this regard.
- It should act as a whole as a club or institution. Environmental sustainability covers a long process. Therefore, all club employees must exhibit a participatory attitude from the top to the bottom. For this, it is very important for club management to include all employees in the process. In this way, all employees will act with the feeling that they are a part of the process. It will be easier to reach the determined targets by carrying out the idea and action processes with a comprehensive participation.
- A unit can be established within the club or institution for the establishment, implementation, monitoring and evaluation of environmental policies. If this is not financially possible, the most suitable candidates among the current employees can be determined and a special working group can be formed, and the process can be managed from a single centre.
- Sports clubs and institutions naturally have external stakeholders from many different sectors. It can establish cooperation with these stakeholders on the environment and its problems. In particular, provisions related to environmental problems can be included in sponsorship agreements that provide a significant economic return. In this way, it can be ensured that the sports club and stakeholders take responsibility. In addition, promotional and educational activities can be created with non-profit institutions in the society. Social responsibility projects can also be carried out with environmental organizations.
- Concrete steps should be taken inside and outside the institution in order to realize the prepared policies and the determined targets. Sustainability studies, which started with small steps, will be a source of motivation to take bigger steps when they start to give results in a short time.
- In order for the environmental policies to be implemented realistically, a certain budget should be created or a share should be allocated from the existing budget, especially in order to take concrete steps. Creating a budget is extremely important for environmental policies, as actions will have a certain cost.

- Training, promotion and awareness activities should be organized under the auspices of sports clubs or institutions. These activities should first start with the employees of the institution, then extend to the fans and finally to the whole society. Due to its nature, sports have the power to affect the masses intensely. Sports clubs and institutions can benefit from this power and increase the awareness of the environment and nature on the society. In addition, it can contribute to providing corporate advantage and increasing social awareness by producing social responsibility projects. Wide participation awareness activities can be organized following the training activities.
- Sports clubs and institutions should establish a feedback mechanism within certain periods. In this way, the steps taken as an organization can be followed and the results obtained can be evaluated. Errors made during the process can be corrected, and existing plans and policies can be improved and developed.
- Finally, in the first step of the process, it is necessary to measure and evaluate the distance covered until the last step. All processes such as activities, collaborations and training activities should be evaluated one by one. This way it can be easier to plan the next step. To what extent were the set goals achieved? What goals were not achieved? A measurement and evaluation system should be established to receive answers to these and similar questions.

The corporate governance model enables institutions and organizations to be managed effectively and efficiently. Institutionalized policies and practices mediate the achievement of an institution's ultimate goals. Today, one of the ultimate goals of sports institutions is the harmony of sports facilities and organizations with the natural environment. Sports institutions are expected to take managerial positions in line with the sustainability goals of the International Olympic Committee. In this section, where the suggestions for environmental sustainability, which is one of the components of corporate governance of sports organizations, are specifically addressed, the roadmap that sports organizations should follow for environmental sustainability is presented as above. The implementation of these suggestions for sports and a sustainable environment is very important for the development of new strategies and sustainable sports and environmental goals. It is inevitable for sports institutions to take a proactive management position in the field of environmental sustainability, to be a role model in the sports industry and to take pioneering steps.

Below is an action plan prepared by researchers for sports clubs and institutions. This action plan outlines the steps to be taken to identify and mitigate negative environmental impacts in sport. This plan can be developed according to the size and sphere of influence of sports clubs and institutions. It summarizes the responsibilities and steps to be taken by the sports industry for a sustainable environment at a basic level.

4.2.1 *An Action Plan Towards Environmental Sustainability in Sports*

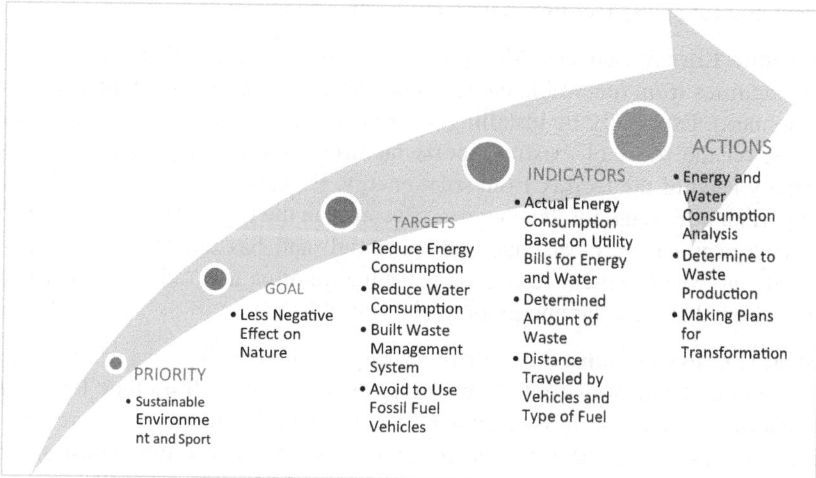

4.3 Third Step Recommendations for Sports Facilities and Organizations

The International Olympic Committee attributes the carbon footprint of sports to two main factors. These are sports organizations and sports facilities. Production, consumption, transportation and accommodation depending on the organizations, and the emergence of air, water and soil pollution depending on the establishment and the energy use of sports facilities [11]. Measures to be taken for sports management aiming at low carbon footprints are grouped under five main headings. These are "*Energy policies*", "*use of environmentally friendly materials*", "*recycling and waste system*" "*waste water harvesting system*" and "*environmentally friendly transportation*" [12].

4.3.1 *Energy Policies (Heating, Cooling and Lighting)*

There is an intense need for energy for heating, cooling and lighting in the facilities that host sports organizations. This need is met from fossil fuel sources. Fossil fuel sources also cause high amounts of carbon dioxide and greenhouse gas emissions.

These emissions have devastating effects on the natural environment, especially the atmosphere, and hinder environmental sustainability targets. These fossil fuel sources are the main reason for the carbon footprint of sports facilities for sustainable environment and sports goals. What measures can be taken within the framework of the energy heading for the construction of an environmentally friendly sports facilities concept and for sports facilities with a low carbon footprint?

Renewable Energy Sources: Meeting the heating, cooling and lighting needs of sports facilities from renewable energy sources (hydraulic, wind, geothermal, solar and biomass). Especially by installing solar energy panels, one of the solar energy systems, on the roofs of existing sports facilities, it seems possible to meet the electricity need of the facility from solar energy. Especially solar energy is seen as an important advantage for the country. "Because, at the beginning of these sources, solar energy is cheap, clean, has high potential, and has almost no harm to the environment. Measurement results on sunshine duration and radiation intensity in Turkey show that the solar energy potential is high" [13].

Central Heating Systems: Since there are large and common areas in sports facilities, the establishment of central heating systems, if any, will provide low energy consumption. Therefore, in cases where it is not possible to convert to renewable energy sources, central heating systems should be installed in the facilities or the existing central systems should be improved to provide optimum efficiency with lower energy.

Energy Saving Practices: Roof, wall and window insulations need to be strengthened in existing sports facilities. Especially in large buildings such as sports facilities, the most important sources of heat loss are doors and windows. Building insulation systems should be made according to climatic and geographical conditions, and this should not be ignored during maintenance and repair processes. In this way, high heat can be provided with less energy. In general, if the necessary importance is not given to thermal insulation in buildings and energy-efficient buildings are not produced, it is clear that energy and environmental problems will be faced in the coming years. While significant economic savings are achieved with the planned insulation works in buildings, negative environmental effects can also be reduced by reducing energy consumption [6].

Installation of LED Lighting Systems: It is necessary to instal LED (Light Emitting Diode) lighting systems in the interior and exterior of the Sports Facilities. LED lighting systems provide optimum lighting by consuming less energy. Therefore, the required lighting can be provided by requiring less energy. "Because LED luminaires are one of the leading lighting systems with rich colour potential, high illumination level, very low energy consumption and extremely economical (up to 90% electricity savings)" [24] (Republic of Turkey Ministry of National Education 2013). "There is now a tremendous shift from traditional lighting technologies (incandescent, fluorescent, high-intensity discharge) to LED lighting. The primary driver of this change has been energy efficiency and the associated cost savings. In addition, LED lighting systems represent an advance in sustainability and it can be said that the advantages will continue to increase as more improvements are made in efficiency" [17].

Motion Sensor Applications in Lighting Systems: Motion-sensitive sensors should be used for lighting systems used especially in multi-purpose sports facilities, inactive sections inside the facility and shower and toilet sections. In this way, unnecessary energy consumption can be prevented during non-use times. "Artificial lighting in buildings is responsible for between 20 and 40% of the world's carbon emissions. Considering that 1 kWh of lighting energy saving will reduce carbon gas emissions by approximately 2.33 kg on an annual basis, the importance of energy saving will be understood even more. Depending on the use of motion-sensitive sensors in buildings, savings of up to 30% can be achieved" [20] (My Sector Magazine 2018). By providing energy savings in this way, besides the economic effects, of course, the negative environmental effects will also be minimized.

Identification of Idle Sections: There are many inactive sections, especially in sports facilities with high usage capacity. These sections, which are expressed as non-usable areas, must be determined by the facility managers. Although it is out of use, energy consumption may also occur in these sections for lighting and heating needs. In order to reduce energy consumption and reduce negative environmental effects, it is very important to determine the idle sections and take the necessary measures.

Landscape Works: The entire environment of the sports facility should be greened with the existing natural vegetation if possible, and the landscaping studies should be designed in accordance with this texture. However, if it is not possible with natural vegetation, it is essential that the surrounding of the building is properly greened and landscaping works are done accordingly. While greening works for the building environment will provide cool air in summer, it will also protect heat in winter. This will minimize the energy need due to electricity consumption. Environmentally friendly landscape works are very common nowadays because environmentally friendly landscape studies enrich the ecosystem in the region, support biodiversity and directly affect the absorption of carbon emissions caused by fuels [7].

Vertical Greening: This method, which is based on the greening of the facade walls of the buildings with certain components and the use of plants, has been frequently applied in recent years in terms of energy saving. It is a method that aims to reduce the electrical energy consumed for cooling the buildings, especially in the summer months due to extreme temperatures. While economic savings are achieved with less energy consumption, the building also becomes a more environmentally friendly structure. In the face of the increasing energy crisis and especially the problems of climate change, designers have started to develop new approaches to reduce the energy requirement in buildings. Among these approaches, façade greening has taken an important place in recent years [7]. Especially vertical greening has become increasingly common in recent years. Because of its positive environmental effects, it is seen as an important solution to climate change problems.

Green Roof Project: Roofs of new or existing sports facilities should be covered with green vegetation. While reducing the effect of solar heat in summer, it has been preferred in recent years as an environmentally friendly practice to prevent heat loss for the building in winter. While the natural environment and biodiversity can be protected thanks to green roof applications, the energy performance of the facility can be increased, the cooling wind effect can be broken, it can act as a filter for particles, carbon dioxide and oxygen balance can be achieved and the service life of the roofs can be extended [3].

4.3.2 *Environmentally Friendly Material Use*

Use of Easily Consumable and Recyclable Products and Materials: It is necessary to use products and materials, also defined as "consumables", in events and organizations held in the sports facility. Especially glass and paper-based products, which can be recycled, are considered environmentally friendly. In this way, environmentally friendly outputs can be achieved and negative environmental effects in waste production are avoided.

Use of Long-Life-Cycle Materials: In the construction and renovation processes, long-lasting materials that will not cause too much construction waste should be used in the facilities. In this way, building systems that can be durable for many years will be preferred and a successful facility can be created in terms of economic and environmental sustainability goals. For example, regardless of its capacity, gyms, etc. Removable and reusable seats can be installed in the fields and halls of the facilities. This kind of application can be applied in fields and halls that require renovation and new seat assembly. "Sustainable building materials, green building materials are materials that do not harm the environment and are used effectively in the production of raw materials sensitive to the limits of exhaustible resources" [23].

Preference of Local Suppliers: Local suppliers should be preferred during the sports organizations and in the procurement of materials from the construction and renovation processes of the facilities. Time and waste can be saved as local suppliers will respond to requests in a short time. In this way, while supporting local development, it is possible to meet the needs of sports organizations and facilities in a short time. In this short time, solid wastes can be minimized and these wastes can be kept under control.

4.3.3 Establishment of Solid Waste (Waste) Management System

Establishment of Waste Separation Units: Sports facilities are used intensively, especially on competition or event days. Due to this mass participation, huge amounts of waste (garbage) are produced. Especially the fact that beverages and foods are in plastic packaging increases the importance of waste separation even more because one of the biggest problems caused by plastic packaging is waste flow. Plastic is a non-renewable material that does not disappear in the environment. The materials used in the production of plastics are cellulose, coal, natural gas, salt and crude oil. Therefore, plastic directly resembles the negative environmental effects of fossil fuels [19]. The collection of these wastes in different garbage containers according to their types can minimize the negative environmental effects of the facilities. In this way, the recycling process of garbage will be accelerated and its reusability will be increased.

Partnerships With Recycling Companies: Collaborations and commercial agreements can be made with companies in the relevant sector in order to recycle the collected wastes by sorting them according to their types. In this way, economic input can be provided for sports facilities within the framework allowed by laws and regulations.

Waste Management Directive: In order to minimize the negative environmental effects caused by wastes, a directive should be prepared describing the process from the collection of waste to recycling. In this way, it can be ensured that the operation is carried out in a balanced way and that the determined goals and objectives are realized. Solid wastes are extremely important due to environmental concerns. Because both directly with the disease-causing or contagious substances in its content; as well as as mice, flies, etc. These substances are carried by pests. Since it is a source of nutrition and reproduction for other living things, it is indirectly an important risk factor for the environment and human health [10].

4.3.4 Rainwater Harvesting System

Rainwater harvesting is the collection and accumulation of rainwater and runoff water, and the provision of water for plant-animal production and domestic consumption [16]. Rainwater harvesting system is of vital importance for sports facilities. There is daily and continuous water consumption, especially for cleaning, showering and landscape irrigation. And in this process, the need for water is met from clean water sources. However, clean water is rapidly depleted due to reasons such as rapid population growth, climate change, environmental pollution and unplanned and uncontrolled water consumption of people [14]. In order to use water resources in a sustainable way and to control the consumption of natural resources in sports facilities, rainwater harvesting systems should be established in the facilities. Harvested rainwater can be used in landscape irrigation, showers and toilets. In this way, clean water resources are also protected. Especially nowadays, with the rainwater collection systems installed on the roofs, the purification and reuse of this water is one of the most environmentally friendly practices.

Controlling Water Consumption: In addition to effective monitoring of water use, low-flow armatures, leak detection systems as well as recyclable water irrigation systems should be installed for landscaping.

4.3.5 Environmental Practices in Transportation

As mentioned in many different points of this study, one of the most important negative environmental outcomes of the sports sector is transportation. Each means of transportation preferred by the team and the fans has environmental impacts of different dimensions. Especially for international sports events and organizations, there is a lot of travel and these travels are usually long distances. Due to these travels, vehicles cause intense carbon emissions. In addition, transportation preferences preferred for participation in local events and organizations organized within the country are another environmental threat. To reduce the carbon footprint from transportation:

- Host country, city and sports clubs can reduce the personal vehicle use of those who come to the facilities by providing a bike sharing service that starts two hours before the start of the competitions and ends two hours later.
- Public transportation (train, bus, etc.) may be free of charge for the participants who prove that they have come to the facilities for the competitions by making the necessary cooperation and joint protocols.
- It may be preferred that the vehicles of the host country, city and sports clubs are electric vehicles. Replacing these vehicles, which are in the inventory of sports clubs, with electric vehicles with environmentally friendly engines can make a very important contribution.
- Awareness of all participating groups should be increased before, during and after travel, and the use of bicycles and electric vehicles as short-distance transportation options should be encouraged.
- Safe walking paths and routes can be created to cover the entire facility and its surroundings in cooperation with public administration units.
- Public awareness can be raised by organizing walking and running events at the local level before the competitions in the countries and cities that host competitions and tournaments [9].

- In many cases, it is accepted that the source of the carbon footprint is individuals. For this reason, carbon footprint reduction or mitigation studies should be designed starting from the individuals rather than the carbon source [18].

References

1. T. Akgeyik, İnsan kaynakları yönetimi boyutuyla kurumsal sosyal sorumluluk: Bir alan araştırması. J. Soc. Policy Conf. **0**(52), 4 (2010)
2. B.Ç. Aksu, A. Doğan, Çevresel sürdürülebilirlik ve insan kaynakları yönetimi fonksiyonlarının yeşil İKY bağlamında değerlendirilmesi. Aksaray Üniversitesi İktisadi ve İdari Bilimler Fakültesi Dergisi **13**(3), 137–148 (2021)
3. İ Ayçam, M. Kınalı, Ofis binalarında yeşil çatıların ısıtma ve soğutma yüklerine olan etkilerinin analizi. Tesisat Mühendisliği Dergisi **135**, 26–34 (2013)
4. M. Boyacı Yıldırım, Nedene dayalı pazarlamada marka-sivil toplum kuruluşu işbirliği: Haytap örneği. Gümüşhane Üniversitesi İletişim Fakültesi Elektronik Dergisi **9**(1), 255–294 (2021)
5. T. Daddi, L. Marrucci, R. Stranieri, D. Casiddu, Evidence based study and action plan to decrease the environmental footprint of grassroots sports: intellectual output 3 sustainable football. EU Green Coach Project Reports (2022)
6. R. Dylewski, J. Adamczyk, Economic and environmental benefits of thermal insulation of building external walls. Build. Environ. **46**(12), 2615–2623 (2011). https://doi.org/10.1016/j.buildenv.2011.06.023
7. E.M. Elgizawy, The effect of green facades in landscape ecology. Procedia Environ. Sci. **34**(2016), 119–130 (2016)
8. C.J. Fombrun, *Reputation: realizing Value from the Corporate Image* (Harvard Business School Press, 1996)
9. L.D. Frank, J.F. Sallis, B.E. Saelens, L. Leary, K. Cain, T.L. Conway, P.M. Hess, The development of a walkability index: application to the neighborhood quality of life study. Br. J. Sports Med. **44**(13), 924–933 (2010)
10. Ç. Güler, Z. Çobanoğlu, *Sağlık açısından çöp, tıbbi dokümantasyon merkezi* (Toplum Sağlığı Dizisi, 1996)
11. International Olimpic Committee, Olimpic documents (2009). https://olympics.com/ioc/documents. Accessed 19 Mar. 2023
12. International Olimpic Committee, Olympic Agenda 2020+515 recommendations (2020). https://stillmedab.olympic.org/media/Document%20Library/OlympicOrg/IOC/What-We-Do/Olympic-agenda/Olympic-Agenda-2020-5-15-recommendations.pdf. Accessed 10 Mar. 2023
13. N. Kaynar, Yenilenebilir enerji kaynaklarından güneş enerjisinin Amasya ilindeki potansiyeli. Bilge Int. J. Sci. Technol. Res. **4**(2), 48–54 (2020)
14. M.Y. Kılıç, M.N. Abuş, Bahçeli bir konut örneğinde yağmur suyu hasadı. Uluslararası Tarım ve Yaban Hayatı Bilimleri Dergisi **4**(2), 209–215 (2018)
15. G.T. Lumpkin, G.G. Dess, Clarifying the entrepreneurial orientation construct and linking it to performance. Acad. Manag. Rev. **21**, 135–172 (1996)
16. T. Oweis, A. Hachum, Water harvesting and supplemental irrigation for improved water productivity of dry farming systems in West Asia and North Africa. Agric. Water Manag. **80**(1–3), 57–73 (2006). https://doi.org/10.1016/j.agwat.2005.07.004
17. P.M. Pattison, M. Hansen, J.Y. Tsao, LED lighting efficacy: status and directionsEfficacité de l'éclairage LED: état de l'art et directions. C R Phys. **19**(3), 134–145 (2018). https://doi.org/10.1016/j.crhy.2017.10.013
18. G.P. Peters, G. Marland, E.G. Hertwich, L. Saikku, A. Rautiainen, P.E. Kauppi, Trade, transport, and sinks extend the carbon dioxide responsibility of countries. Clim. Change **97**, 379–388 (2009)

19. K. Sabaliauskaitė, D. Kliaugaitė, Resource efficiency and carbon footprint minimization in manufacture of plastic products. Environ. Res. Eng. Manag. **1**(67), 25–34 (2014)
20. My Sector Magazine, Hareket Sensörü İle Enerji Tasarrufu Nasıl Sağlanır? (2018). https://www.sektorumdergisi.com/hareket-sensoru-ile-enerji-tasarrufu/. Accessed 9 Mar. 2023
21. A. Sevim, S. Bülbül, UEFA Finansal Fair Play Kriterleri kapsamında Türk futbolunda finansal raporlamanın önemi ve bir sistem gerekliliği. Kara Harp Okulu Bilim Dergisi **27**(2), 187–212 (2018)
22. The Roadmap for Creating a Sustainability Strategy was developed within the framework of the As Sustainable As Possible (ASAP) project co-funded by the Erasmus + Programme of the European Union
23. M.Z. Tufan, C. Özel, Sürdürülebilirlik kavramı ve yapı malzemeleri için sürdürülebilirlik kriterleri. Int. J. Sustain. Eng. Technol. **2**(1), 9–13 (2018)
24. Republic of Turkey Ministry of National Education (2013). http://megep.meb.gov.tr/mte_program_modul/moduller_pdf/Aydınlatma%20Elemanları.pdf. Accesseed 9 Mar. 2023
25. J.C. Villar, S.L. Hidalgo, A.C. Penela, B.G. Meijide, A new perspective for labeling the carbon footprint against climate change. Global Warming—Impacts and Future Perspective (2012). https://doi.org/10.5772/46533
26. O. Yaşar, N. Çantay, Kurumsal itibar algısında örgütsel iletişimin rolü: Devlet ve özel okullarında karşılaştırmalı bir analiz. OPUS Int. J. Soc. Res. **14**(20), 1161–1186 (2019)